David M. Hata

Upper Saddle River, New Jersey
Columbus, Ohio

Library of Congress Cataloging-in-Publication Data

Hata, David M.
Introduction to vaccum technology / David Hata.
p. cm.
ISBN-13: 978-0-13-045018-0
ISBN-10: 0-13-045018-9
1. Vacuum technology. I. Title.
TJ940.H384 2008
621.5'5--dc22
2006038591

Editor in Chief: Vernon Anthony
Editor: Jeff Riley
Editorial Assistant: Lara Dimmick
Production Editor: Stephen C. Robb
Design Coordinator: Diane Y. Ernsberger
Cover Designer: Terry Rohrbach
Cover art: Index Stock
Scanning Coordinator: Karen L. Bretz
Scanning Technician: Janet Portisch
Production Coordination: PineTree Composition, Inc.
Production Manager: Matt Ottenweller
Director of Marketing: David Gesell
Marketing Manager: Ben Leonard
Marketing Assistant: Les Roberts

This book was set in Times Roman by Pine Tree Composition, Inc. It was printed and bound by R. R. Donnelley & Sons, Inc. The cover was printed by R. R. Donnelley & Sons, Inc.

Pearson Education Ltd.
Pearson Education Singapore Pte. Ltd.
Pearson Education Canada, Ltd.
Pearson Education—Japan
Pearson Education Australia Pty. Limited
Pearson Education North Asia Ltd.
Pearson Educación de Mexico, S.A. de C.V.
Pearson Education Malaysia Pte. Ltd.

10 9 8 7 6 5 4 3 2 1
ISBN-13: 978-0-13-045018-0
ISBN-10: 0-13-045018-9

PREFACE

This book is written for users of vacuum systems, especially the technicians who are responsible for maintaining them. It is also written for instructors who, like me, have been frustrated with the lack of suitable teaching materials for a technician-level course taught at community colleges. For these instructors, suggested laboratory exercises have been included to support a lecture-laboratory course.

Vacuum systems are critical to many industries. They are vital to establishing required process pressures, establishing a clean process environment, and removing reaction byproducts from the process chamber. Often vacuum systems are taken for granted until there is a malfunction that changes the process environment, resulting in improperly processed products, and ultimately, a loss of revenue for the company.

Most vacuum books approach the subject topically. That is, separate chapters are devoted to rough vacuum pumps, high-vacuum pumps, pressure gauges, vacuum materials, and other topics. This organization is appropriate for a reference book but not for a teaching text.

This book approaches vacuum systems from a pressure regime viewpoint. That is, after covering some basic chemistry, the first pressure regime covered is the rough vacuum regime. Within the study of rough vacuum systems, the following topics are covered: gas load, pumping mechanism, pressure measurement, and vacuum system construction. The discussion of rough vacuum is then followed by the study of high-vacuum systems. The same topics are revisited, but this time from a high-vacuum perspective. Once both rough vacuum and high-vacuum systems are covered, then the topics of leak detection and residual gas analysis are introduced.

This pedagogical approach lends itself to laboratory experimentation. During the review of gas laws from chemistry, there are experiments and demonstrations that can be performed to reinforce basic laws and concepts. Then, during the study of rough vacuum systems, pump-down times can be calculated and pump downs performed in the laboratory. Likewise, during the study of high-vacuum systems, pump downs as well as other lab exercises, such as outgassing and residual gas analysis, can be performed.

My intent was not to create an exhaustive treatment on vacuum science. Some topics have been consciously omitted, such as ultrahigh vacuum systems and the pumps that are used to create these very low pressures. The question driving decisions to include or exclude material in the book focused on the needs of technicians in a production environment and the types of vacuum systems used.

This book is the result of years of teaching vacuum systems courses at the community college level. I thank my former students for challenging me to learn more about vacuum science and systems as we worked to understand the behavior of our vacuum systems. I enjoyed working with them in the vacuum laboratory and experiencing their joy of discovery. It is because of them that I was motivated to write this book.

Acknowledgments

Writing a textbook is truly a team effort. It takes many people with many talents to produce a textbook.

I would like to express my appreciation and thanks to all of you who played a part in publishing this textbook. First of all, I thank Jeff Riley and Steve Robb at Pearson Education, Inc. I also thank John Shannon at Pine Tree Composition for overseeing production, and Melissa Akers, MOA Editorial Services, for helping secure the needed permissions. Thanks to all who helped with the editing, typesetting, graphics, and other production steps.

I also thank all the individuals who helped secure permission to use figures from their company publications: Steve Hansen at MKS, Inc., David Ames at Stanford Research Systems, Betty Ann Kram at Inficon, Michele Haurin at Scott Specialty Gases, Laura McNally at BOC Edwards, Richard Rauth at KNF Neuberger, Steve Palmer at Varian, and Eric Houge at Pfeiffer Vacuum.

Finally, I especially thank the following reviewers for their invaluable input during development of the manuscript: Marilyn Barger, Hillsborough Community College; Jim Jozwiak, Boise State University; and Stephen Hansen.

CONTENTS

Introduction to Vacuum Technology

CHAPTER 1

Much Ado about Vacuum

1.1 INTRODUCTION

A new episode of *ER* had just begun when Tom's beeper sounded. "Now what?" he mouthed as he reluctantly reached for his cellphone and keyed in the number on his pager.

Back at the fabrication plant (known as simply "the fab"), the night shift had been struggling to keep one of the plasma etchers on-line, but increasing pump-down times pointed to a problem that could no longer be ignored.

Tom would be in the fab in twenty minutes, but it would be another two hours before a worn valve seat would be diagnosed as the culprit and the valve replaced. By then, *ER* would be long over, but it had been captured on videotape. Tomorrow was Tom's day off, and he would view the episode after a good night's sleep.

Sound like fiction? Maybe, maybe not. To many people, "vacuum" is synonymous with "Hoover." To Tom and others, vacuum technology is a vital part of the manufacturing processes, and we, as end users, enjoy the benefits of the products they produce.

Take an ordinary lightbulb, for example. Once the filament, base, and bulb are made, they are fitted together by automated manufacturing systems. First, the filament is mounted on the stem assembly and the ends of the filament are clamped to the two lead-in wires. Next, most of the air inside the bulb is pumped out, creating a vacuum, and the bulb is filled with a mixture of argon and nitrogen. Finally, the base, bulb, and stem assembly are joined and sealed. The finished bulbs are tested, and those that pass the final test are packaged and shipped.

The chief problem that slowed the development of a commercial incandescent lightbulb was finding suitable filament materials. In the early 1800s, Sir Humphrey Davy found that platinum was the only metal that could produce white heat for any length of time. Many metals that were tried had failed because they quickly oxidized in air, causing the filament to burn out. The solution to the oxidation problem was to create a vacuum that would keep air (which is 20% oxygen) from reaching the filament material, thus preserving the light-producing material.

Thomas Alva Edison is credited as the inventor of the first commercially practical electric lamp in 1879. His lamp used carbonized cotton threads as the filament material. The new lamp burned for two days and forty minutes before it burned out on October 21, 1879, the usual date given for Edison's invention.

Of course, Edison's original design has undergone numerous changes and refinements. Nevertheless, today's incandescent lightbulbs closely resemble Edison's original lamps. The major difference is that tungsten is the filament material of choice for today's bulbs, and different gases are used for higher efficiency. The point is that without vacuum technology, we would not have incandescent lightbulbs, fluorescent lamps, tungsten halogen lamps, mercury vapor lamps, neon lamps, or metal halide lamps. We might still be living and working by the light from candles and oil lamps.

There are many other examples that demonstrate how crucial vacuum technology is in the manufacture of many products today, but we need to move on. For more information on the history of vacuum technology and how it is used in manufacturing processes visit the Web sites for the American Vacuum Society (www.avs.org) and the Society of Vacuum Coaters (www.svc.org) .

1.2 WHAT IS A VACUUM?

You probably have a notion of the meaning of the term vacuum. As a noun, it may refer to a vacuum cleaner or some other contraption that sucks up objects. In another sense, one might say that when you leave a group, your departure creates a "vacuum." So with all these uses of the term, it makes sense to define how we intend to use it in our discussions ahead.

In the ideal sense, a vacuum is a space that contains absolutely nothing, a space totally devoid of matter. However, in actual practice, such a perfect vacuum is impossible, even for the best of vacuum systems. Even under the best of vacuum conditions, a chamber still contains hundreds of thousands of gas molecules per cubic centimeter.

A better definition of *vacuum,* for the purposes of this text, might be a space partially emptied (to the highest degree possible) by artificial means (using a pump). This results in a volume containing far fewer gas molecules than the same volume in the surrounding atmosphere, that is, a degree of *rarefaction* below atmospheric pressure. If the surrounding atmosphere is at normal atmospheric pressure, then a vacuum exists if the density of the gas molecules in the chamber is less than the density of the gas molecules outside the chamber.

TABLE 1.1
Pressure regimes

Name of pressure range	Pressure in torr
Rough vacuum	759 to 1×10^{-3}
High vacuum	1×10^{-3} to 1×10^{-8}
Ultrahigh vacuum	1×10^{-8} to 1×10^{-16}

Using this definition, vacuum describes pressures that range over nineteen orders of magnitude, from 1 atmosphere (760 torr) to 1×10^{-16} torr. Often, this range of pressures is arbitrarily divided into regimes or regions that are given the following labels: rough vacuum, high vacuum, and ultrahigh vacuum. Table 1.1 gives one way of defining these pressure regimes. Each *pressure regime* has its own unique characteristics and requires its own types of pumps and pressure gauges. All of this will be covered in greater detail in the coming chapters.

1.3 BENEFITS OF A VACUUM

Pundits predicted that the advent of the transistor and the decline in the use of vacuum tubes in radios, televisions, and similar instruments would result in a decline in the vacuum industry—and maybe its demise. However, a funny thing happened along the way. Instead of dying, the vacuum equipment industry has grown in parallel with the microprocessor, memory, and application-specific segments of the semiconductor industry and the even larger surface-coating industry, all of which rely on vacuum technology.

Vacuum technology is used extensively throughout the wafer fabrication plant, and many steps in the manufacture of integrated circuits require vacuum systems to create the proper process environment before chemical and physical manufacturing processes can take place. These process steps include etching surface layers, depositing materials on the wafer surface, and cleaning wafer surfaces.

The health of the vacuum industry is evident in the number of suppliers of vacuum equipment at industry trade shows such as SEMICON®West and annual conferences of the American Vacuum Society and the Society of Vacuum Coaters. Those numerous equipment vendors, together with the many companies that integrate vacuum subsystems into their products, results in a very large industry segment that is essential for the manufacture of the products we use, enjoy, and often take for granted.

Creating a vacuum simply involves any process that removes gas molecules from the gas phase within a closed space, enclosure, or process chamber so that there are fewer gas molecules per unit volume inside than the number of gas molecules in the same volume outside. There are two basic ways to do this. One method uses a vacuum pump to capture gas molecules from the chamber and exhaust them to a system that neutralizes any harmful or toxic gases. If no harmful gases are present in the effluent, the gas can be exhausted directly into the atmosphere.

Another method of creating a vacuum changes the state of the gas molecules within the pump. In other words, it removes the gas molecules from the gas phase. This change of state can be accomplished by condensing the gas molecules or turning them into a solid through a chemical reaction with highly reactive materials. The resulting product then sticks to the interior of the pump.

Why go to the trouble and expense of creating a vacuum? This is an important question that must be answered before we begin our study of vacuum systems and related topics. Creating a vacuum can be costly and at times frustrating, but for many products, creating a vacuum is a requisite to carrying out manufacturing processes. Figure 1.1 lists some of the benefits of a vacuum environment.

Removing gas molecules from the enclosure or process chamber achieves several objectives. First, it results in a cleaner process environment in the chamber by removing potential contaminants. Contaminants can be particles as well as unwanted gases that, if allowed to remain in the chamber, can participate in the reactions taking place in the chamber and produce unwanted compounds.

Second, creating a vacuum in the chamber increases the mean free path of the gas molecules. The *mean free path* of a gas molecule is the average distance the gas molecule will travel before colliding with another gas molecule. As the chamber pressure is reduced, the gas molecules in the chamber become more widely spaced. Thus, they are able to travel farther before colliding with another gas molecule. This is important in manufacturing processes like *sputtering,* in which argon atoms are ionized and accelerated toward a *target* to sputter off atoms from the target, and in *ion implantation,* in which an ion beam is directed at the wafer. A high molecular density of gas molecules would impede the movement of metal atoms to the wafer surface in the sputtering process and would hinder the ion beam from reaching the wafer in the ion implantation process.

Third, creating a vacuum is a way of controlling the number of collisions of molecules against a surface. This is important in sputtering metal layers and in controlling the rate of film growth in *chemical vapor deposition.*

Fourth, creating a vacuum lowers the *molecular density.* Again, consider the case of a lightbulb. If the inside of the lightbulb were not evacuated to remove air, any oxygen molecules left in the bulb when the filament is heated would react with the metal in the filament, and the metal filament would burn up. Lower molecular density also aids in freeze-drying. When the saturated layer of gas over an evaporating surface is removed, a much higher evaporation rate can be achieved without having to raise the ambient temperature. Lower molecular density can reduce heat conduction. A good example is the heat conduction between the inner and outer walls of a thermos bottle or vacuum dewar.

FIGURE 1.1
Benefits of a vacuum.

Produces a cleaner environment
Increases mean free path
Controls the number of surface collisions
Lowers molecular density
Creates a force
Reduces heat flow
Increases vaporization
Protects materials

Fifth, creating a vacuum can result in a force that is useful in the manufacturing process. This force is used in vacuum wands to hold the wafer to the wand as they are manually moved from one place to another or in vacuum chucks to hold wafers in place during a manufacturing process. It can also be used in dynamic applications to move solid or liquid material through a pipe or duct where the higher atmospheric pressure drives the material toward the lower pressure in an attempt to equalize the pressure at both ends of the pipe or duct.

Sixth, creating a vacuum reduces *heat flow*. If the pressure is reduced, the distance between gas molecules increases and the number of molecular collisions will lessen, resulting in a reduction of heat flow. Changing the gas can also reduce heat flow if the new gas has a lower heat conductivity.

Seventh, creating a vacuum can increase vaporization. *Vaporization* occurs when liquid molecules leave a surface and do not return. It works like this: If the chamber containing the liquid is evacuated, fewer molecules will be above the surface of the liquid. Thus, a vaporizing molecule is less likely to collide with a gas molecule in the vapor phase because there are fewer molecules to hit, bigger spaces between the molecules, and fewer molecules impacting the liquid surface.

Finally, creating a vacuum can protect materials from reaction with air. To accomplish this, the chamber is first evacuated and then backfilled with an inert gas in order to remove as many reactive gas molecules as the process dictates.

1.4 PROCESSES THAT USE A VACUUM

Table 1.2 lists some of the processes that are carried out in each of the three pressure regimes. This is by no means an exhaustive list, but it does give a few examples for each regime.

In the *rough vacuum* regime, foods can be sealed under vacuum to remove air from containers. An example is Astronaut Ice Cream, a freeze-dried product that resembles and tastes like ice cream and is a favorite food of the author's wife. Semiconductor manufacturing processes that are carried out in this pressure regime include sputtering and holding wafers on chucks.

High vacuum is required in heat treating, decorative coating, and chemistry research. Semiconductor manufacturing processes requiring high vacuum include chemical and

TABLE 1.2
Processes that use vacuum

Rough vacuum	High Vacuum	Ultrahigh vacuum
Food processing	Heat treating	Materials research
Freeze drying	Vapor deposition	Metallurgy
Evaporation	Ion implantation	Surface analysis
Sputtering	Surface coating	Molecular beam epitaxy
Distillation	Thermal insulation	Physics research
Electric conduction	Chemistry research	Space research

physical vapor deposition, etching, and ion implantation. All of these processes require a greater degree of rarefaction.

Ultrahigh vacuum is used primarily in research studies, such as materials research, physics research, and surface analysis. It is not used in many manufacturing processes because of the time it takes to reach pressures in the ultrahigh-vacuum regime and the cost to maintain them.

1.5 CREATING A VACUUM

There are several ways to create a vacuum in an enclosed space. Probably the simplest and least expensive way is to increase the volume of the enclosed space. Glass handles like the ones used to move sheets of plate glass exploit this technique. The handles have a rubber diaphragm that is placed against the glass surface. A mechanism deforms the rubber diaphragm, pulling it away from the glass surface. This forms an enclosed space with a lower gas density than the surrounding atmosphere. For this to work, the seal at the edge of the rubber diaphragm must be airtight.

Another way to create a vacuum is to lower the temperature of the gas inside an enclosed volume. As the temperature is lowered, the gas molecules travel slower. When they hit a surface, the force they exert on it is smaller. Hence, the force per unit area, or pressure, is lower.

If the temperature is low enough, the gas molecules can be frozen. Freezing gas molecules takes them out of the gas phase by changing the gas to a solid. This reduces the density of molecules in the gas phase and thus lowers the pressure. Cyrogenic pumps, or cryo pumps, for short, work on this principle.

Another method of creating a vacuum by changing the gas to a solid involves a chemically reactive material. In this type of pump, the gas molecules react with a material like titanium to form a solid compound. The resulting solid then sticks to the interior surfaces of the pump.

A more common and more efficient way of creating a vacuum is to displace mechanically the gases from an enclosed space or chamber. To accomplish this, most vacuum applications use a single or multiple pumps to remove gas from a process chamber. Rotary vane, scroll, diaphragm, screw, and blower-type pumps are used as means to create a vacuum.

Each of these pumping mechanisms leads to different pump designs, offering options in vacuum systems design. Some of these different pump designs will be described in chapters 4 and 5.

1.6 LOOKING AHEAD

The material presented in this book is not intended to be an exhaustive treatment of vacuum systems. Instead, it provides a fundamental understanding of vacuum systems for technicians and others who will be working with vacuum systems on a regular basis.

Our study of vacuum systems will begin with a review of the behavior of gases in chapter 2. You probably have studied the behavior of gases in a general chemistry course, so the material in this chapter will be familiar. For a more in-depth treatment, any general chemistry text would be a good reference on this topic.

Chapter 3 will introduce vacuum system components. The presentation, by design, will be general. Chapters 4 and 5 will then provide a more detailed discussion of the vacuum systems used in the rough and high-vacuum regimes. The chapter organization will flow from a discussion of gas load to pumping mechanisms and pressure measurement techniques to vacuum system troubleshooting.

Chapter 6 will focus on a new measurement instrument, the *residual gas analyzer*. Unlike pressure gauges, the residual gas analyzer provides partial pressure information for each gas being pumped.

A discussion of leak detection follows in chapter 7. Unfortunately, leaks occur, and in the case of vacuum systems, gas leaks provide a way for gas outside the system to enter.

Chapter 8 provides an introduction to gas delivery and pressure control. Mass flow controllers will be described, along with several pressure control strategies. Finally, chapter 9 will conclude our discussion with safety issues related to vacuum systems.

BIBLIOGRAPHY

Danielson, Phil. "Why Create a Vacuum?" *Vacuum & Thinfilm,* July 1999, pp. 18–23.

Quirk, Michael, and Julian Serda. *Semicoductor Manufacturing Technology.* Prentice-Hall, Upper Saddle River, NJ, 2001.

Tompkins, Harland G. *The Fundamentals of Vacuum Technology,* 3rd ed. AVS Monograph Press, New York, NY, 1997.

Varian Vacuum Products. *Product Catalog 2000,* Varian Vacuum Technologies, Lexington, MA.

PROBLEMS

Using the Internet or other references, answer the following questions.

1. Who invented the vacuum cleaner?
2. Name the company that built the first electric vacuum cleaner that used both a cloth filter bag and cleaning attachments. When did this happen?
3. Who invented the first practical electron tube? When? What was the electron tube called?
4. Who invented the triode tube? When?
5. What company was the first to manufacture a commercial electron tube?
6. Who were the following persons, and what part did each play in the history of vacuum science?
   a. Evangelista Torricelli
   b. Robert Boyle
   c. Otto von Guericke
   d. Ludwig Eduard Boltzmann
   e. Jacques-Alexandre Charles
   f. John Dalton

CHAPTER 2

The Behavior of Gases

2.1 INTRODUCTION

The key to understanding the operation of vacuum systems is the ability to visualize how gases behave. This is where our discussion of vacuum systems begins. Some of this information will be familiar to you, because living on Earth teaches us many things about the behavior of gases. For example, we have observed the results of the movement of gas molecules on a windy day and enjoyed the work of an electric fan on a hot day. We have seen vapor rising from a lake in the early morning. We have observed that gases are different. A helium balloon, for instance, rises in air, whereas balloons filled with air at room temperature rest on the floor. We have also seen carbon dioxide gas from dry ice cascading over the edge of a punch bowl.

We have used pumps to concentrate gas in a confined space, as when we pump up the tires of a bicycle or car. We have enjoyed the benefits of a reduction in gas molecules between the walls of a Thermos bottle that slows the movement of heat and keeps liquid refreshments hot or cold. We have observed the results of matter changing from a gaseous state to a liquid state (*condensation*), or from a solid state to a gaseous state (*sublimation* of carbon dioxide). These are only a few examples, and you can probably think of many more.

The goal of this chapter is to review the laws of chemistry and physics that govern the behavior of gases. The material in this chapter will help you develop a mental picture, or model, of the behavior of gases so that you will be able to visualize what is actually happening inside a vacuum system.

2.2 STATES OF MATTER

Before we begin our study of gases, let us review what we know about matter. Matter can exist in four states: solid, liquid, gas, and plasma. Matter in its lowest energy state exists as a *solid.* Take water, for example. Water exists in the solid state as ice at temperatures below 0°C at or near sea level. The water molecules in ice are held in a rigid structure that gives the ice shape. When a piece of ice is placed in a container, the ice will partially fill the container but will not take the shape of the container.

When the temperature is increased above the freezing point, water changes from the solid state to the liquid state. Water exists as a *liquid* between the temperature of 0°C and 100°C. A liquid can flow and can conform to the shape of the containers into which it is placed. For example, water placed in a pitcher conforms to the pitcher's shape. At the surface of the water is an *interface* between the water below and the air above. At this water-air interface, water molecules, if they have enough energy, can escape from the liquid state. This process is called *evaporation.* Conversely, water molecules in the vapor phase can enter the liquid. We know this process as *condensation*.

At 100°C, the boiling point of water is reached. The temperature of the water will remain at the boiling point until all of the water is converted to a *vapor.* When in the gaseous state, the water vapor (steam) can now fill the entire space that encloses it. In the gaseous state, water molecules are able to move independently about the space. It is the *gas state* of water vapor and other gases that we will study in the chapters that follow.

In the last state of matter, *plasma,* the gas molecules are ionized, forming a mixture of *particles*. There are *electrons*, *ions*, *radicals*, *neutrals*, and other energetic particles. Plasmas hold the promise of energy generation through *fusion* and are used in myriad manufacturing processes that produce products we enjoy every day. This is a fascinating state of matter, but unfortunately the topic is beyond the scope of this book.

Looking at the *periodic table of elements,* illustrated in Figure 2.1, we find elements that exist naturally in all three states. Most of the elements exist at room temperature as

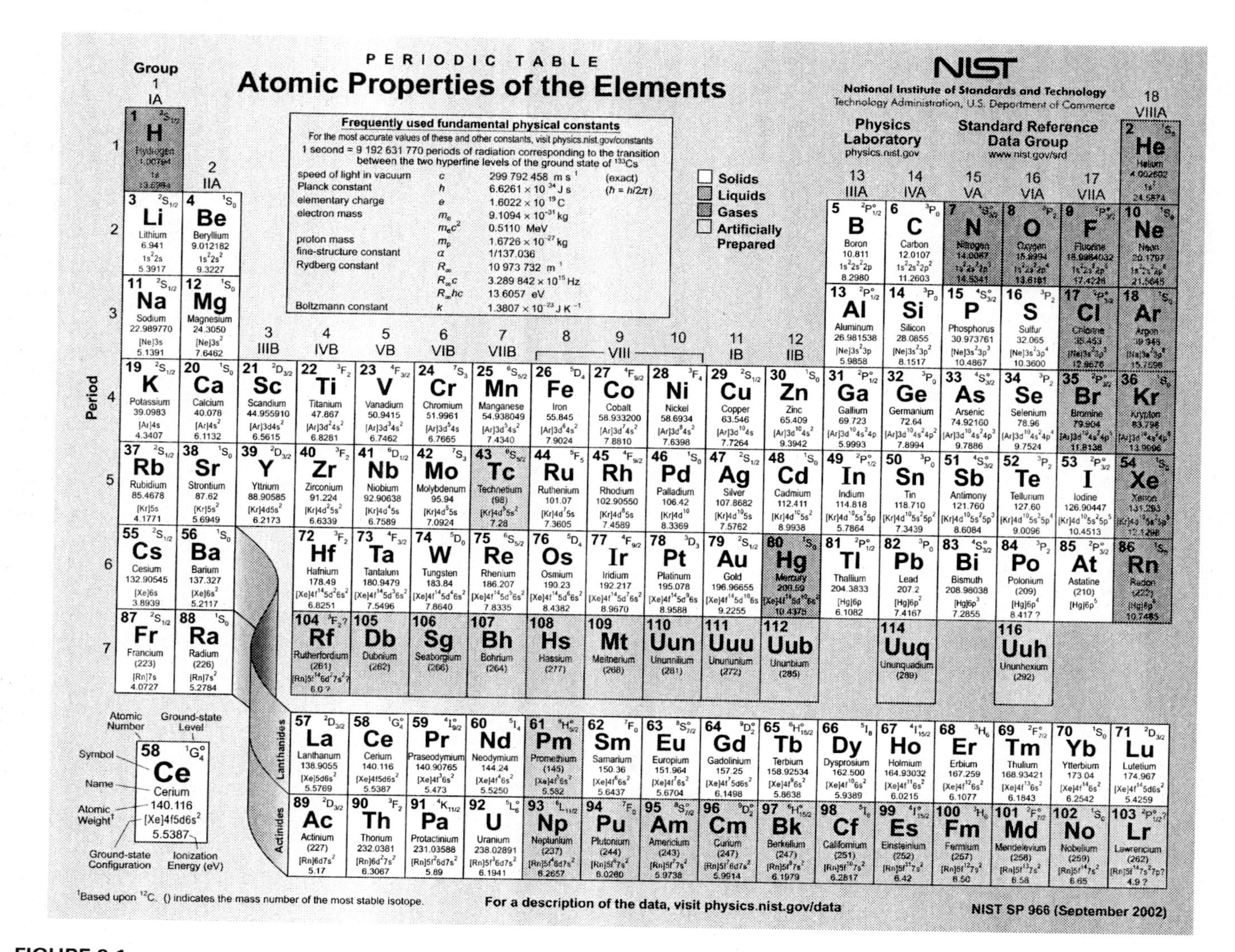

FIGURE 2.1
The Periodic Table of Elements.

solids: copper, iron, and silicon are examples. Only one element, mercury, exists at room temperature as liquid. The remaining elements exist as gases at room temperature, for example nitrogen, oxygen, and argon.

With this brief review of the states of matter, let us begin our study of gases. We will start by developing a mental picture of how gases behave. Our mental picture will be based on the kinetic theory of gases.

2.3 KINETIC THEORY OF GASES

The *kinetic theory of gases* attempts to explain the physical behavior of gases. As is typical of theories in science, it is based on a set of assumptions. In the case of the kinetic theory of gases, the underlying assumptions include the following:

- Gases are composed of a very large number of separate, independent atoms and molecules that are in constant, straight-line motion. These molecules are so widely spaced that, on the average, the total volume of the atoms and molecules is very small compared to the total volume of the gas. In the absence of collisions, the gas atoms and molecules display no attraction or repulsion toward one another.
- The straight-line motion of the gas molecules may be interrupted by collisions with other gas molecules or with the walls of the chamber. These collisions are assumed to be completely elastic, resulting in no energy being converted to heat or another form of energy or a change in the atoms or molecules involved in the collision.
- Because it is in constant motion, each gas molecule possesses a certain amount of *kinetic energy*. Since the gas molecules can have different velocities, each gas molecule can have a different kinetic energy. The total kinetic energy possessed by a gas is a function of the temperature of the gas and the number of gas atoms or molecules in the chamber. The average kinetic energy of the gas atoms and molecules is directly proportional to the absolute, or *Kelvin* temperature of the gas.

Sometimes it is helpful to try to draw a picture to help visualize something. Draw an enclosed space of any shape that we will call a chamber. Inside this chamber, draw some small circles. These small circles will represent the gas molecules. They should be randomly placed throughout the chamber. To represent the motion of the gas molecules, draw an arrow from each gas molecule to represent its direction of travel. Use the length of the arrow to represent the speed of the molecule—a longer arrow can represent a high speed of travel, and a short arrow represent a low speed of travel. Some gas molecules will collide with each other, and some will collide with the walls of the chamber. How can you represent these events? Is your drawing similar to the one shown in Figure 2.2?

How could you modify your drawing to represent a mixture of two different gases such as helium and nitrogen? How would you draw the air that we breathe? Take a few moments to consider these scenarios before continuing in this chapter.

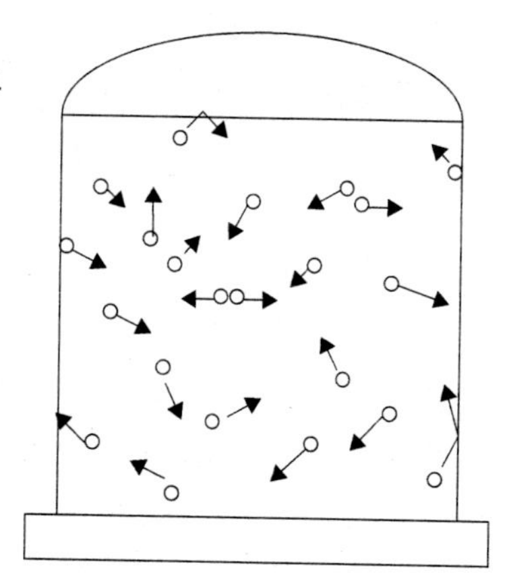

FIGURE 2.2
Gas molecules within a process chamber.

The kinetic theory of gases can be used to explain the following properties of gases:

- Gases fill a container or chamber completely, irrespective of pressure and the shape of the container.
- Gases are compressible (*Boyle's law*).
- Gases expand/contract with changes in temperature (*Charles' law*).
- Gases of equal volume and pressure have an equal number of molecules (*Avogadro's law*).
- The total pressure of a mixture of gases is equal to the sum of the partial pressures of each gas in the mixture (*Dalton's law*).

Let us now use the kinetic theory of gases to explain these observed properties of gases.

2.3.1 Distribution of Gas Molecules in an Enclosed Volume

When a certain amount of gas is placed within an enclosed volume or chamber, the gas fills the entire volume and the density of gas molecules will be uniform throughout the volume. If this were not true, there would be areas of greater concentration of gas molecules and areas of lesser concentration. This would produce a *flux* that would cause the gas molecules to move from areas of higher concentration to areas of lesser concentration. This movement of gas molecules would continue until the concentration difference and the flux are reduced to zero. Hence, the density of gas molecules must be uniform throughout the entire enclosed volume.

The density of the gas molecules varies with pressure. The relationship between pressure and molecular density is shown in Table 2.1.

At 760 *torr*, there are approximately 3×10^{19} gas molecules per cubic centimeter, about 30 million trillion. This is a lot of gas molecules, and they are constantly bumping into each other and against the walls of the chamber. They are so close to one another that, on the average, they travel only 6.4×10^{-5} mm between collisions.

At one *millitorr*, a reduction of six orders of magnitude from atmospheric pressure, the number of gas molecules per cubic centimeter is approximately 4×10^{13} or 40 trillion. This still seems like a lot of gas molecules, but at this pressure, the distance traveled between collisions

TABLE 2.1
Molecular density as a function of pressure.

Pressure in torr	Molecules per cubic centimeter
760	3×10^{19}
1	4×10^{16}
1×10^{-3}	4×10^{13}
1×10^{-6}	4×10^{10}
1×10^{-9}	4×10^{7}

approaches 2 in. (5.1 cm). In large chambers, the number of gas molecule to gas molecule collisions dominates, whereas there are proportionately few collisions against chamber walls.

At a pressure of 1×10^{-9} torr, another reduction of six orders of magnitude, the number of gas molecules per cubic centimeter is now 4×10^{7}, or 40 million. That is still a lot of gas molecules in a small space, but now the distance between the collision of two gas molecules is 31 mi (50 km), a distance much greater than between the walls of the chamber. At these pressures, the collisions between the gas molecules and the chamber walls dominate. This change from gas molecule to gas molecule collisions to gas molecule to chamber wall collisions is important in processes that are carried out at these low pressures.

2.3.2 Compressibility of Gases

One of the qualities that distinguishes gases from solids and liquids is *compressibility.* From our life experiences, we know that gases can be compressed. For example, we have compressed a gas when pumping up a tire.

The compressibility of gases can be explained quite easily using the kinetic theory of gases. Gas molecules are seen as very small molecules separated by great distances. Reducing the volume enclosing the gas molecules crowds them together. As the gas molecules fly about in the smaller volume, the number of collisions between them increases, as does the number of collisions between the gas molecules and the chamber walls. The cumulative effect of all of the collisions between the gas molecules and the wall of the chamber is a greater force on the chamber wall, or *chamber pressure,* as shown in Figure 2.3.

The relationship between the volume of a gas and the pressure exerted by a constant amount of gas is stated in Boyle's law. *Boyle's law* states that if the temperature and the amount of the gas are held constant, the volume of the gas is inversely proportional to the pressure exerted by the gas. This relationship can be expressed in equation form as

$$V \propto \frac{1}{P} \qquad \text{(temperature is held constant)} \qquad \textbf{(2.1)}$$

The proportionality can be changed to an equality by inserting a proportionality constant, *k*. Hence,

$$V = k \times \frac{1}{P} \qquad \text{(temperature is held constant)} \qquad \textbf{(2.2)}$$

If a fixed volume of gas at a constant temperature is compressed or expanded, then

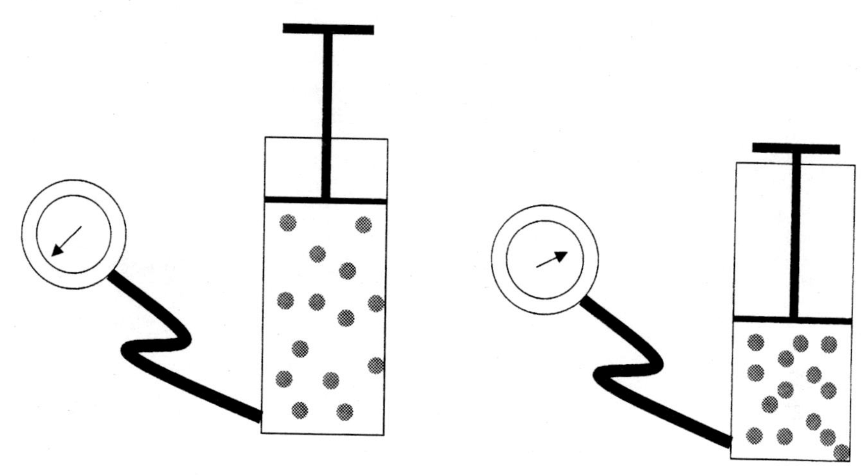

FIGURE 2.3
Pressure-volume relationship.
Pushing down on the handle decreases the volume of the gas and increases its pressure if the air cannot escape and the volume is constant.
Source: Brady, *General Chemistry,* Fig. 11.6, p. 336.

$$P_i \times V_i = k \qquad \text{where } P_i \text{ and } V_i \text{ are the initial conditions,}$$

and

$$P_f \times V_f = k \qquad \text{where } P_f \text{ and } V_f \text{ are the final conditions.}$$

Since the two equations equal the same constant, we can set an equals sign between them:

$$P_i \times V_i = P_f \times V_f \tag{2.3}$$

This is the mathematical statement of Boyle's law.

If the amount of gas and the temperature are constant, we can use Boyle's law to solve for the missing parameter if the other three parameters are known. For example, if we know the initial pressure and volume and the final volume, we can solve the equation $P_i \times V_i = P_f \times V_f$, for P_f. This yields

$$P_f = \frac{P_i \cdot V_i}{V_f}. \tag{2.4}$$

EXAMPLE 2.1 A gas is compressed at a constant temperature from an initial volume of 1.5 L to 1.0 L. If the initial pressure is 800 torr, what is the final pressure in torr?

Solution

From Boyle's law,

$$P_i V_i = P_f V_f = \text{constant}$$

where P_i = the initial pressure,
P_f = the final pressure,
V_i = the initial volume,
and V_f = the final volume.

Solving for P_f,

$$P_f = \frac{P_i \cdot V_i}{V_f}$$

$$P_f = \frac{(800 \text{ torr})(1.5\text{L})}{1.0\text{L}}$$

$$P_f = 1200 \text{ torr}$$

2.3.3 Effects of Temperature

Since gas molecules are in constant motion, each gas molecule has a certain amount of kinetic energy. The kinetic energy is equal to $\frac{1}{2}mv^2$, where m is the mass of the gas molecule and v is the velocity, or speed, of the molecule. The gas molecules in the chamber do not all have the same kinetic energy. The distribution of kinetic energies follows a curve known as the *Maxwell-Boltzmann distribution.*

Figure 2.4 shows the velocity distribution for four gases at the same temperature. Examining the graph, we see that the number of gas molecules with zero kinetic energy is essentially zero because few, if any, gas molecules are standing still at any instant. As we move to the right along the horizontal axis, the kinetic energy increases, as does the fraction

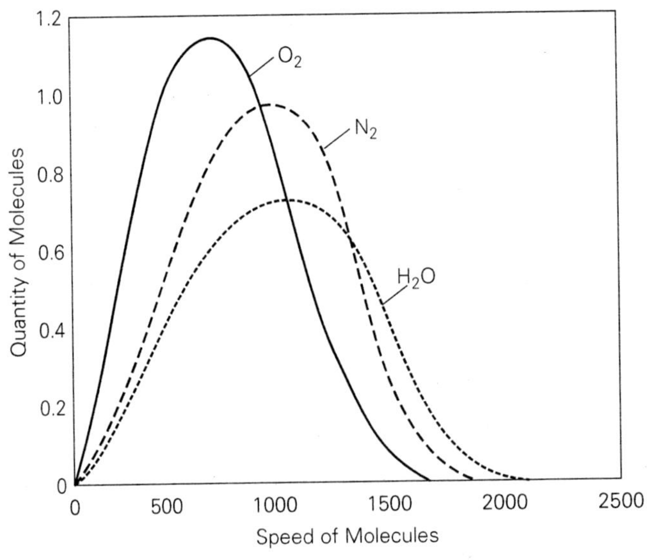

FIGURE 2.4
Distribution of velocities for four gases at the same temperature.
Source: Tocci and Viehland, *Chemistry: Visualizing Matter,* Fig. 10.8, p. 361.

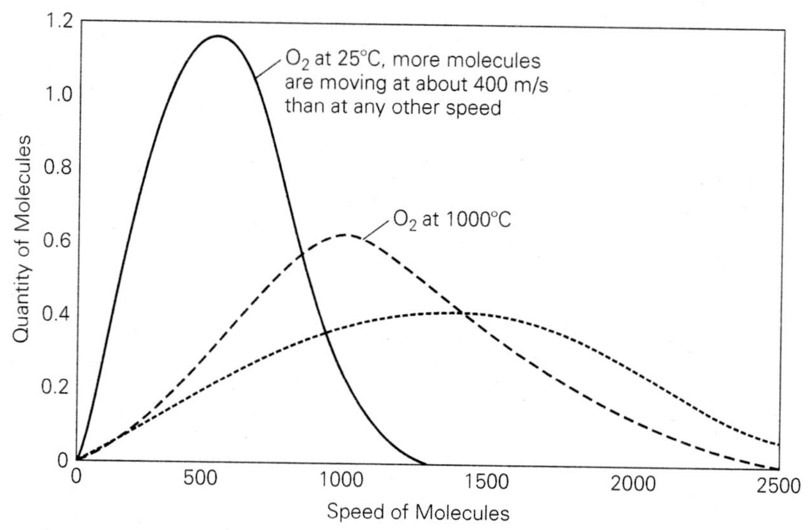

FIGURE 2.5
Velocity distribution for the same gas at different temperatures.
Source: Tocci and Viehland, *Chemistry: Visualizing Matter,* Fig. 10.9, p. 362.

of gas molecules possessing each kinetic energy. The graph peaks at a certain kinetic energy called the *most probable kinetic energy*—that is, the most frequently found kinetic energy if we were to examine the gas molecules at a specific point in time. From the peak, as the kinetic energy increases, the fraction of gas molecules possessing each kinetic energy begins to decrease and approaches the horizontal axis. The average kinetic energy of the gas is higher than the most probable kinetic energy because the curve is not symmetric.

When the temperature of a gas changes, the shape of the curve described by the *Maxwell-Boltzmann distribution* changes. For example, if the temperature of the gas is increased, the kinetic energy of the gas molecules will also increase. The shape of the curve flattens out, and more gas molecules are found at higher kinetic energies. Corresponding increases are seen in both the most probable kinetic energy and the average kinetic energy. Figure 2.5 shows the effect of increasing or decreasing temperatures on the velocity distribution for a gas.

The relationship between volume and temperature, at a constant pressure, is given by Charles' law, which is named after eighteenth-century French chemist Jacques Alexander Charles. *Charles' law* states that the volume of a gas is directly proportional to the absolute temperature at a constant pressure. Mathematically, Charles' law can be expressed as

$$V \alpha T \qquad \text{(at a constant pressure)} \tag{2.5}$$

where V is the volume of the gas and T the absolute temperature. Charles' law can also be expressed as an equality:

$$V / T = k = \text{constant.} \tag{2.6}$$

It is important to note that when using the equation for Charles' law, temperatures must be in units of *Kelvin.* That is, temperature given in Celsius or Fahrenheit must be first converted to Kelvin.

EXAMPLE 2.2 A certain amount of gas occupies a volume of 3.0 L at a temperature of 30°C and a pressure of 1 atm. How many liters will it occupy at a temperature of 75°C and a pressure of 1 atm?

Solution

From the data given, the pressure remains constant, while volume and temperature change. According to Charles' law,

$$\frac{V_i}{T_i} = \frac{V_f}{T_f} \tag{2.7}$$

Solving for V_f,

$$V_f = V_i\left(\frac{T_f}{T_i}\right)$$

$$V_f = (3.0\ \text{L})\left(\frac{75 + 273\text{K}}{30 + 273\text{K}}\right)$$

$$V_f = (3.0\ \text{L})\left(\frac{348\text{K}}{303\text{K}}\right)$$

$$V_f = 3.45\ \text{L}$$

And finally, *Gay-Lussac's law* describes how the pressure of a gas depends on its temperature while the volume of the gas is held constant. According to this law, the pressure of a gas at a constant volume is directly proportional to the absolute temperature. Stated in equation form,

$$P \,\alpha\, T \qquad \text{(at a constant volume)} \tag{2.8}$$

or using a proportionality constant,

$$\frac{P}{T} = k = \text{constant} \qquad \text{(at a constant volume)}. \tag{2.9}$$

EXAMPLE 2.3 A gas exerts a pressure of 500 torr at 25°C. How many torr will the gas exert if the temperature is raised to 40°C? Assume that the volume remains constant.

Solution

From Gay-Lussac's law,

$$\frac{P_i}{T_i} = \frac{P_f}{T_f} \tag{2.10}$$

Solving for P_f,

$$P_f = P_i \frac{T_f}{T_i}$$

The temperatures, given in Celsius, must be converted to absolute temperature in degrees Kelvin.

$$T_i = 25°\text{C} = 25 + 273\text{K} = 298\text{K}$$
$$T_f = 40°\text{C} = 40 + 273\text{K} = 313\text{K}$$

Substituting for P_i, T_i, and T_f in the P_f equation,

$$P_f = (500 \text{ torr})\left(\frac{313\text{K}}{298\text{K}}\right)$$

$$P_f \approx 525 \text{ torr}$$

2.3.4 Gases of Equal Volume and Pressure

Amadeo Avogadro proposed *Avogadro's law,* which states that equal volumes of gases at the same temperature and pressure contain equal numbers of gas molecules. Avogadro's law was subsequently used to show that gases such as hydrogen, oxygen, and chlorine must be *diatomic* in structure.

If we examine one mole of gas at a temperature of 0°C, or 273 K, and a pressure of 1 atm, or 760 torr, we can measure the molar volume at *STP* (standard temperature and pressure). Table 2.2 gives the molar volume for some common gases. From the results of these measurements plus measurements for other gases, the average volume occupied by one mole of gas at STP is 22.4 liters, and this value is therefore taken to be the molar volume of an ideal gas at STP.

TABLE 2.2
Molar volumes for some common gases.

Gas, symbol	Molar volume in liters
Argon, Ar	22.397
Carbon dioxide, CO_2	22.260
Helium, He	22.434
Hydrogen, H	22.433
Nitrogen, N_2	22.402
Oxygen, O_2	22.397

Source: Adapted from Brady, *General Chemistry,* Table 11.1, p. 343.

TABLE 2.3
Gas composition of air.

Gas	Symbol	Percent by volume	Partial pressure
Nitrogen	N_2	78	593 torr
Oxygen	O_2	21	158 torr
Argon	Ar	0.94	7.1 torr
Carbon dioxide	CO_2	0.03	0.25 torr
Helium	He	0.0005	4×10^{-3} torr
Hydrogen	H_2	0.00005	4×10^{-3} torr
Water	H_2O	Variable	5–50 torr

2.3.5 Mixtures of Gases

When two or more gases that do not react chemically are placed in the same confined space, the pressure exerted by each gas in the mixture is the same as it would be if it were the only gas in the confined space. The pressure exerted by each gas in a mixture is called its *partial pressure.* The *total pressure* is equal to the sum of the partial pressures of each gas in the mixture. This statement is called *Dalton's law* of partial pressure. Mathematically, it can be expressed as,

$$P_{\text{total}} = P_{\text{gasA}} + P_{\text{gasB}} + P_{\text{gas C}} + \ldots. \quad \textbf{(2.11)}$$

where P_{total} = the total pressure of the gas mixture,
P_{gasA} = the partial pressure of gas A,
P_{gasB} = the partial pressure of gas B,
and so on.

Consider the composition of our atmosphere. Table 2.3 lists the gases that make up our atmosphere, the percentage by volume, and the partial pressure of each constituent gas.

If we add all of the partial pressures for the constituent gases that make up air, the sum of the partial pressures is approximately 760 torr.

2.4 IDEAL GAS LAW

The three gas laws explained in Section 2.3 can be combined into one law known as the *combined gas law.* The combined gas law can be expressed as

$$\frac{PV}{T} = \text{constant} \quad \textbf{(2.12)}$$

As with Boyle's law, Charles' law, and Gay-Lussac's law, the amount of gas is assumed to remain constant.

The combined gas law allows us to analyze situations where two of the three parameters change at the same time. The following example illustrates the use of the combined gas law.

EXAMPLE 2.4 A sample of gas, stored in a 0.5-L container at 25°C, exerts a pressure of 750 torr. What would be the pressure of the gas if the gas sample was transferred to a 1-L container at 50°C?

Solution

Let us begin by organizing the given information in tabular form as shown below.

	Initial condition	Final conditions
Pressure, P	750 torr	?
Volume, V	0.5 L	1.0 L
Temperature, T	25 + 273 K	50 + 273 K

Solving the combined gas law equation for the final pressure, P_f, yields

$$P_f = P_I\left(\frac{V_i}{V_f}\right)\left(\frac{T_f}{T_i}\right)$$

$$P_f = (750 \text{ torr})\left(\frac{0.5 \text{ L}}{1.0 \text{ L}}\right)\left(\frac{323\text{K}}{298\text{K}}\right)$$

$$P_f = 406 \text{ torr}$$

If we allow the amount of gas to vary, then the constant in the combined gas law will also vary. For example, if the number of moles of gas is doubled, the value of the constant will double. If the number of moles of gas is reduced by one half, the value of the constant in the combined gas law will likewise be reduced by one half.

If n represents the number of moles of gas, then the combined gas law can be written as

$$\frac{PV}{T} = n \times (\text{other constant}) \tag{2.13}$$

Many chemistry textbooks use the letter R to represent this other constant and name it the *universal gas constant.* Hence, the equation can be written as

$$PV = nRT \tag{2.14}$$

Equation 2.14 is also known as the *ideal gas law.* Using the ideal gas law, if three of the variables are known, then the fourth can be found.

The value of the universal gas constant, R, can be found experimentally by finding the volume of one mole of gas under a known condition. One condition that can be used is STP. The conditions for STP are 0°C (273 K) and 1 atm (760 torr). The volume of one mole of gas varies somewhat under these conditions. For example, the molar volume at STP of oxygen is 22.397 liters, nitrogen is 22.402 liters, hydrogen is 22.433 liters, and carbon dioxide is 22.260 liters. Averaging the molar volumes of many gases yields an average volume occupied by one mole of gas at STP of 22.4 liters.

From the ideal gas law,

$$R = \left(\frac{PV}{nT}\right)$$

$$= \left(\frac{mL - torr}{mol - K}\right)$$

$$R = 0.0821\left(\frac{L \cdot atm}{mol \cdot K}\right)$$

The universal gas constant can be expressed using other units, e.g., milliliters, torr, moles, and degree K. The conversion can be performed as follows,

$$R = \left(0.0821 \frac{\text{L} \cdot \text{atm}}{\text{mol} \cdot \text{K}}\right)\left(1000 \frac{\text{mL}}{\text{L}}\right)\left(760 \frac{\text{torr}}{\text{atm}}\right)$$

$$R = 6.24 \times 10^4 \frac{\text{mL} \cdot \text{torr}}{\text{mol} \cdot \text{K}}$$

EXAMPLE 2.5 What volume will 24.0 g of nitrogen (N_2) occupy at 20°C and a pressure of 0.755 atm?

Solution

Algebraically solving the ideal gas law for volume, *V*, yields

$$V = \frac{nRT}{P}.$$

Since pressure is given in atmospheres, we can use the gas law constant 0.0821 L · atm mol^{-1} · K^{-1} and convert the given data to the units used in this constant. Hence,

$$n = 24\text{g of } N_2 \times \frac{1 \text{ mol of } N_2}{28\text{g of } N_2} = 0.86 \text{ mol of } N_2$$

$$T = 20°\text{C} + 273 = 293\text{K}$$

Substituting these values into the equation for volume yields,

$$V = \frac{(0.86 \text{ mol})\left(0.0821 \frac{\text{L} \cdot \text{atm}}{\text{mol} \cdot \text{K}}\right)(293\text{K})}{0.755 \text{ atm}}$$

$$V = 27.4 \text{ L}$$

2.5 GAS PRESSURE

A gas is made up of many gas molecules in rapid motion. Inside a closed chamber, the gas molecules collide with each other and with the chamber walls. Each collision with the chamber wall exerts a small force on the chamber wall. The cumulative effect of all the molecule–wall collisions can be expressed as a force per unit area or pressure, such as pounds per square inch.

$$\text{Pressure} = \frac{\textit{force exerted by gas molecules}}{\textit{unit area of surface}}$$

There are a great many units of measure used to specify the pressure of a gas. In the United States, the most familiar units are pounds per square inch (psi) and inches of mercury. For example, the air pressure in the tires of your car is specified in psi. Also, when weather reporters tell you the atmospheric pressure reading in your geographic area, they will give you the atmospheric pressure in inches of mercury, although the units will usually be omitted.

Other pressure units include atmosphere, torr, bar, and Pascal. The pressure unit, *torr*, is often used in the United States and is the standard unit of pressure used in this textbook. In parts of the world that have adopted the *Sysétme international des unites* (SI), the standard unit of pressure is *newton per square meter* (N/m^2) or *Pascal* (Pa), although *millibar* (mbar) is used as well. It is important for you to be able to convert pressure units. The following pressures are equivalent to 1atm of pressure:

$$\begin{aligned}1 \text{ atm} &= 760 \text{ mm of mercury (Hg)}\\ &= 760 \text{ torr}\\ &= 101{,}325 \text{ Pa } (\text{N}/\text{m}^{-2})\\ &= 101.325 \text{ kPa}\\ &= 1.01325 \text{ bar}\\ &= 1013.25 \text{ mbar}\end{aligned}$$

Some books on vacuum technology give conversion tables for changing pressure in one unit to an equivalent pressure in another unit measure. This is helpful in situations in which you happen to have the table with you or have committed the table to memory. Another method that depends less on memory comes from dimensional analysis. It is based on the mathematical fact that multiplying the original pressure by the number 1 does not change the value of the original pressure. The trick is to find the right form of the number 1. The form of 1 that should be used has 1 atm expressed in the new unit of pressure in the numerator and 1 atm expressed in the original unit of pressure in the denominator. The original pressure is then multiplied by this ratio. Mathematically, it can be expressed as follows:

$$P_{new} = P_{original}\frac{(1 \text{ atm})_{new}}{(1 \text{ atm})_{original}}. \tag{2.15}$$

The following example illustrates how using our method adapted from dimensional analysis works. Note that the units should cancel, leaving only the new unit of pressure.

EXAMPLE 2.6 Convert 100 torr to Pascals.

Solution

To convert pressure in torr to an equivalent pressure in Pascals, we need to know, from memory or a table, the number of each of these units in 1 atm.

1 atm = 760 torr
1 atm = 101,325 Pascals

The pressure unit torr is our original pressure, and the pressure unit Pascal (Pa) is our new pressure unit. Hence,

$$P_{new} = P_{original}\frac{(1 \text{ atm})_{new}}{(1 \text{ atm})_{original}}$$

$$P_{new} = 100 \text{ torr}\frac{101{,}325 \text{ Pa}}{760 \text{ torr}}$$

$$P_{new} = 13{,}300 \text{ Pa}$$

Despite the myriad of pressure units, one would think that measuring pressure in a chamber would be no more complicated than measuring temperature. That there is a variety of pressure gauges seems logical, because thermometers, too, are made to measure different temperature ranges, such as outdoor thermometers, meat thermometers, candy thermometers, and are calibrated in different units, including degrees Fahrenheit and degrees Celsius. However, as we will see, accurately measuring the pressure of a gas is not a trivial task, and is certainly more complex than measuring temperature.

In the rough vacuum regime, there are basically two types of pressure gauges: indirect reading pressure gauges and direct reading gauges. *Indirect reading gauges* use the heat-conduction property of gases to determine the density of the gas and thus the pressure in the chamber. The more gas molecules, the greater the heat loss, and thus the higher the pressure. *Thermocouple* and *Pirani* pressure gauges are examples of indirect reading gauges.

Direct reading gauges measure the force exerted on a surface by the gas molecules. In a *capacitance manometer,* a thin diaphragm is deformed by the force exerted by the gas molecules. This deformation changes the capacitance between the diaphragm and a reference electrode. The gauge controller then converts this change in capacitance into a pressure reading. A *Bourdon gauge,* on the other hand, uses a curved copper tube to sense the difference between atmospheric pressure on the outside and chamber pressure on the inside to generate the mechanical force that moves the indicator needle.

In the high-vacuum regime, neither mechanical force nor the heat transfer mechanism can be used to measure pressure. To measure pressures in the high-vacuum regime, gas molecules are converted from neutral atoms and molecules to ions. The ions are then collected, and the resulting current is measured and converted to a pressure reading.

Because the heat-conduction properties of various gases differ, and so does the ease of ionization, the pressure readings obtained will have to be adjusted depending on the gas or gas mixture. This makes pressure measurement more complex than just reading a number on the pressure readout or display.

2.6 MEAN FREE PATH

In a process chamber at atmospheric pressure, the gas molecules are constantly colliding with one another and with the walls of the chamber. The distance traveled by a given gas molecule between two successive collisions is called the *free path.* The distance a gas molecule travels and the direction of travel are randomly distributed. Hence, the *mean free path* is the average of the free paths traveled by the gas molecules in the chamber. According to kinetic theory, this can be represented mathematically as

$$\lambda = \frac{1}{\sqrt{2}\pi d_o^2 n} \quad \textbf{(2.16)}$$

where λ is the mean of the free paths, d_o is the molecular diameter in meters, and n is the gas density in molecules per cubic meter.

EXAMPLE 2.7 Use the equation for λ to estimate the mean free path of air molecules at 760 torr and 0°C.

Solution

From a table of physical properties of gases, the molecular diameter of air molecules is 0.372 nm at $T = 0°\text{C}$. The density of air at 760 torr is approximately 3×10^{25} molecules per cubic meter. Hence,

$$\lambda = \frac{1}{\sqrt{2}\pi(0.372 \times 10^{-9}\ \text{m})^2(3 \times 10^{25}\ \text{molecules/m}^3)}$$

$$\lambda = 5.42 \times 10^{-6}\ \text{cm}$$

The mean free path of the gas molecules is related to gas density and thus to the pressure of the gas within the chamber. The lower the pressure, the lower the density and the greater the spacing between gas molecules. In this case, because the gas molecules are more widely spaced, a gas molecule will travel a greater distance before colliding with another gas molecule. Conversely, a higher pressure means a greater density of gas molecules and a shorter distance between collisions, that is a shorter mean free path.

If temperature is relatively constant, a rule-of-thumb relationship between mean free path and gas pressure can be described mathematically as

$$\lambda = \frac{5 \times 10^{-3}\ \text{torr} \cdot \text{cm}}{P} \tag{2.17}$$

where λ is in centimeters, and P is in torr.

EXAMPLE 2.8 Use Equation 2.17 to find the approximate mean free path of air at 760 torr.

Solution

Substituting 760 torr into Equation 2.17 yields

$$\lambda = \frac{5 \times 10^{-3}\ \text{torr} \cdot \text{cm}}{760\ \text{torr}}$$

$$\lambda = 6.58 \times 10^{-6}\ \text{cm}$$

Note that the result obtained using Equation 2.17 is close, but not exactly equal, to the value we obtained earlier using Equation 2.16. That is OK. The important idea here is the inverse relationship between mean free path and the pressure. The mean free path and the pressure are inversely related. As the pressure goes down, the mean free path goes up, and conversely, as pressure goes up, the mean free path goes down.

Why is the mean free path of atoms and molecules an important parameter in manufacturing processes? An example is *sputtering* of metal atoms to form a metal film on the surface of a silicon wafer. In this process, metal atoms are sputtered, or knocked off, a metal target. The metal atoms must travel from the target to the silicon wafer, located a certain distance away. If the mean free path is too short, the metal atoms will be unable to travel unimpeded from the target to the wafer surface. If the sputtered metal atoms collide with other atoms or molecules, they might change direction and miss the wafer altogether, or they might lose so much kinetic energy that they will be unable to bond to the wafer surface. In either case, the rate of growth of the metal film will be affected, and it will take longer to achieve the desired thickness of the metal film.

2.7 ADSORPTION AND DESORPTION

Adsorption is a process whereby something sticks to the surface of something else. In a vacuum system, that "something" is usually water vapor, because water vapor, in the form of humidity, is always found in air. The "surface" is the interior surface of the process chamber and the surface of the wafer being put into the chamber.

Any material exposed to air for any length of time will come to *equilibrium* with the surrounding water vapor—that is, a point where the number of water molecules attaching to the surface will equal the number of water molecules leaving the surface. At this point, the surface is said to be *saturated.*

Since gas molecules are in constant motion, some of the water molecules in the gas phase will come into contact with surfaces in the chamber per unit time. When the water molecules hit the surface, some of them attach to the surface and are said to be *adsorbed* on the surface. The attachment occurs because there are stable positions for them on the surface. In addition, the impinging atoms or molecules must give up enough of their kinetic energy to become trapped on the surface. Sometimes, the impinging atoms or molecules do not lose the required amount of kinetic energy and are quickly reflected from the surface. The *trapping, or sticking, coefficient* ranges from 0.001 to 0.9 for various gases and substrates. As the water molecules are adsorbed on the surface, they are instantly replaced in the air above the surface by other water molecules because of the constant interdiffusion produced by molecular motion.

The polar nature of water molecules does not simply cause them to adhere to the chamber surfaces; they also adhere to each other in ever-weakening bonds as the layers of adsorbed water molecules become thicker and thicker and more disordered. Over time, the adsorbed water molecules can form many monolayers of water molecules on surfaces. Figure 2.6 shows the hydrogen bonding between water molecules.

As a pump-down cycle proceeds to pressures below 1 millitorr, the adsorbed water molecules begin to *desorb,* that is, leave the surface to which they have been bonded. In fact, water vapor becomes the dominant gas source and determines the gas load for the vacuum system. The weakly held outer layers will desorb quickly, but the more tightly held layers will take longer to desorb.

Desorption of water from a surface can be speeded up by adding energy to the surface and the water molecules. One method is to heat the surface with a radiant energy source such as a heat lamp. Energy from the heat source provides the energy for the water molecules to break the *hydrogen bonds* that hold them to each other. Another method is to

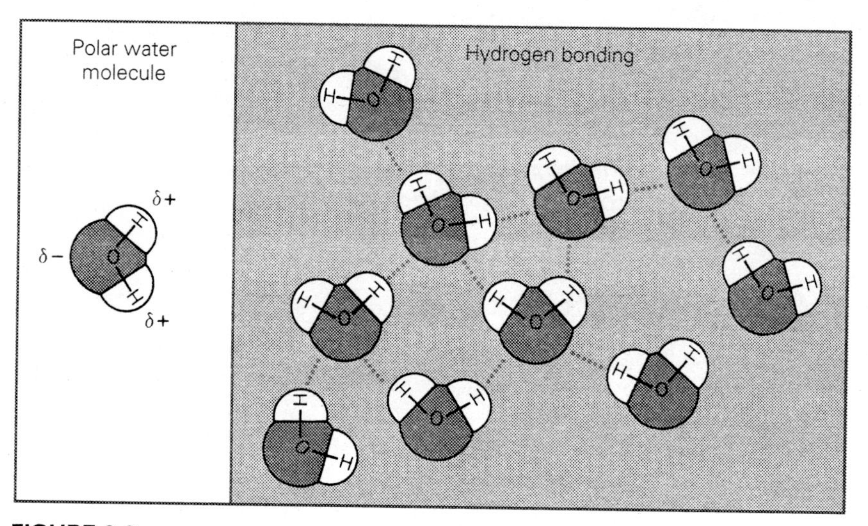

FIGURE 2.6
Hydrogen bonding between water molecules.
Source: Brady, *General Chemistry,* Fig. 12.2, p. 368.

put vacuum components into an oven to drive the water molecules from the surface. This method is called *bake out.*

Purging the chamber with a dry gas will also remove water adhering to the interior surface of the chamber. A gas often used for this purpose is dry nitrogen. The nitrogen gas molecules strike the surface and dislodge water molecules.

2.8 DIFFUSION AND PERMEATION

Diffusion and permeation are often confused. Let us see if we can differentiate the two terms.

Diffusion is the movement of molecules within a material, whether it be a solid, liquid, or gas. Diffusion is the process by which atoms or molecules move from an area of high concentration to an area of low concentration within the material. Perfume diffuses in the air in a room. A drop of ink diffuses when dropped into a beaker of water.

Permeation, on the other hand, is a three-step process whereby a gas moves through a material, typically a solid. First, gas adsorbs on the outer wall of the material, then it diffuses through the bulk material, and last, it desorbs from the inner surface. In a vacuum system, steady-state permeation acts like a constant leak.

2.9 THERMAL CONDUCTIVITY

The ability of various gases to conduct heat is not the same. Using the data found in Appendix B.2 of *A User's Guide to Vacuum Technology,* by John F. O'Hanlon, the *thermal conductivity* of air is 24.0 mJ/(s K), while argon has a thermal conductivity of 16.6 mJ/(s K). This says that air is a much better conductor of heat than argon. On the other hand, the thermal conductivity of helium is 142.0 mJ/(s K), a value about six times greater than the thermal conductivity of air.

The thermal conductivity of gases is relatively constant in the rough vacuum regime, but as the pressure approaches 10^{-2} torr, the thermal conductivity of a gas approaches zero. We will return to this gas parameter when we discuss pressure gauges that rely on the thermal conductivity of gases for their operation.

2.10 VAPOR PRESSURE

When a liquid is placed in a closed container or chamber, the molecules that have enough energy to escape the liquid surface and accumulate in the chamber. Once in the vapor phase, they exert a pressure, just like other gas molecules in the chamber. The pressure will rise initially and then gradually becomes constant, having reached an *equilibrium vapor pressure.*

This behavior of vapor pressure can be visualized in the following manner. Upon introduction of a liquid into the chamber, molecules begin to evaporate, moving from the liquid state to the vapor phase. Once in the vapor phase, the evaporated molecules begin to exert a pressure against the walls of the chamber. As time passes, more and more molecules move to the vapor phase. At the same time, some of these molecules collide with the liquid surface where they transfer some of their energy back to the molecules in the liquid. Lacking enough energy to return to the vapor phase, these molecules once again become part of the liquid. This is the process of *condensation.* Hence, there are two opposing processes working at the same time, evaporation of molecules from the surface and condensation of molecules back to the liquid. When these two opposing processes are equal, the equilibrium vapor pressure has been reached, as depicted in Figure 2.7.

Vapor pressure is a function of temperature. The relationship between vapor pressure and temperature can be described by the following mathematical equation:

$$\log_{10} P_{vp} = A - \frac{B}{T} \tag{2.18}$$

where P_{vp} is the vapor pressure in torr,
A and B are constants for a specific gas,
and T is the temperature in degrees Kelvin.

If temperature were to increase, then the negative term in the equation becomes smaller and the right-hand side of the equation becomes a larger number. Hence, the logarithm of the vapor pressure becomes larger, and the vapor pressure becomes greater. On the other hand, if the temperature decreases, the negative term increases, and when subtracted from A makes the value of the right side of the equation smaller. If the logarithm of the vapor pressure is smaller, then the vapor pressure must decrease.

What happens when the vapor pressure of the liquid equals the pressure above the liquid? We have all observed what happens in cooking food. When water in a pan is heated to a high-enough temperature, it boils. Raising the temperature of the water increases the vapor pressure of the water. The boiling point is therefore the temperature at which the

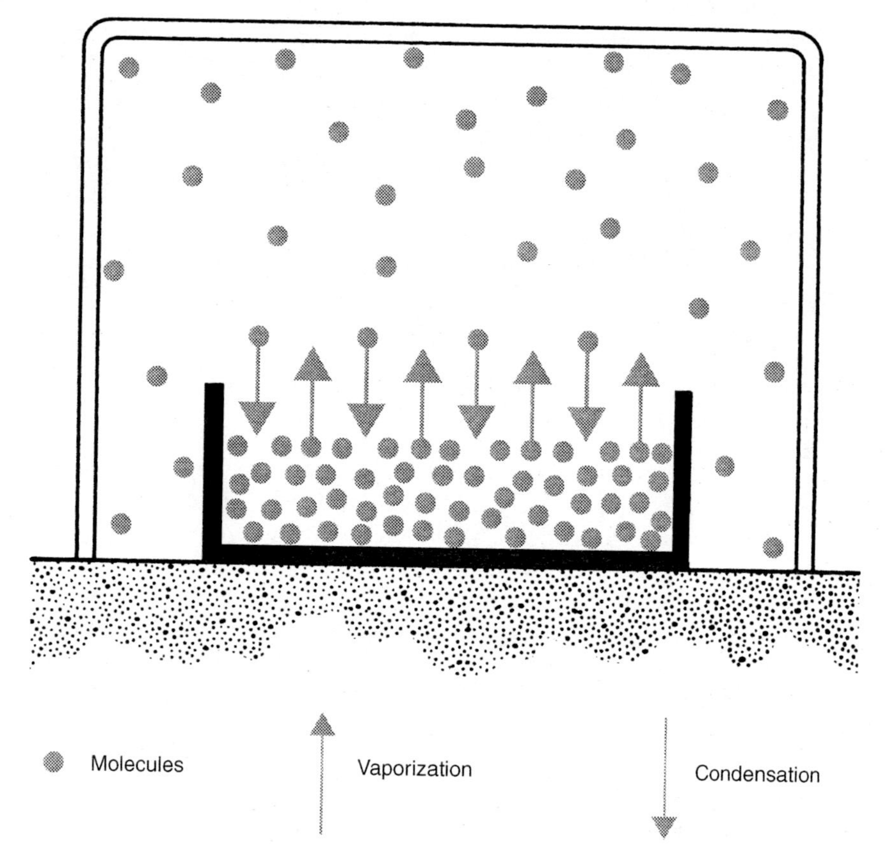

FIGURE 2.7
Equilibrium is reached in the chamber. Rates of evaporation and condensation are equal.
Source: Brady, *General Chemistry,* Fig. 12.15, p. 379.

vapor pressure of the water equals the atmospheric pressure. At 760 torr, water boils at 100°C, and this temperature is referred to as the *normal boiling point* of water. At higher atmospheric pressures, the boiling point of water will be higher; conversely, at lower atmospheric pressures, the boiling point will be lower.

What happens if the temperature of a liquid is increased? The highest temperature at which a distinct liquid phase can exist is called the *critical temperature,* and the associated vapor pressure is called the *critical pressure.*

Solids can also evaporate just like liquids. However, in the case of solids, the process is called *sublimation.* Dry ice is a familiar example. In dry ice, a crystalline solid, the molecules vibrate about their equilibrium positions and undergo collisions with their nearest neighbors. This gives rise to a distribution of energies, similar to a liquid. At the surface of the solid, some of the molecules possess kinetic energies large enough to enable them to break away from the surface and enter the vapor phase. If this process occurs in a closed chamber, an equilibrium condition will occur when the rate at which molecules leave the solid will equal the rate at which they return to the solid. The pressure exerted by the vapor is the *equilibrium vapor pressure* of the solid.

A commercial example that uses sublimation is freeze-drying. Freeze-dried instant coffee begins as a batch of brewed coffee. The brewed coffee is then frozen and the ice (water) component is removed under vacuum. In this way, the delicate heat-sensitive molecules are preserved and the flavor is retained.

SUMMARY

Knowledge of the behavior of gases is useful in understanding the operation of vacuum systems. Some key points follow:

- Gases will expand to fill the enclosure in which they are confined.
- The kinetic molecular theory is a model of gas behavior.
- Pressure, temperature, volume, and the number of moles of a gas are four variables that define a gaseous system. The ideal gas law relates these four parameters.
- Gases can move via the processes of diffusion and permeation.
- Gases can adsorb and desorb from surfaces.
- Gases can conduct thermal energy.
- Gas pressure can be given in a number of different units.

BIBLIOGRAPHY

LeMay, H. Eugene, Jr., et al. *Chemistry: Connections to Our Changing World.* Upper Saddle River, NJ: Prentice-Hall, Inc., 1996.

Brady, James E. *General Chemistry: Principles & Structure,* 5th ed., Hoboken, NJ: John Wiley, 1990.

Mahan, J. E. *Physical Vapor Deposition of Thin Films.* Hoboken, NJ: John Wiley, 2000.

O'Hanlon, J. F. *A User's Guide to Vacuum Technology,* 3rd ed., Hoboken, NJ: Wiley Interscience, 2003.

Tocci, S. and C. Viehland. *Chemistry: Visualizing Matter.* Fort Worth, TX: Holt, Rinehart & Winston, 1996.

PROBLEMS

1. Using the concepts presented in this chapter, explain why:
   a. It is more difficult to open the freezer door just after you have opened it only a few moments before.
   b. A balloon filled with helium is buoyant one day, but it has less "lift" the next day.

c. It is difficult to separate two plastic cups when one is stacked inside the other for more compact storage.
d. A bag of potato chips seems to have more air in it while in flight than on the ground at the airport.

2. In our local newspaper the other day, the barometic pressure was given as 29.93" and falling. In what units is this barometic pressure reading given?
3. A gas is compressed from a volume of 0.60 liters to 0.35 liters. If the initial pressure was 500 torr, what is the final pressure? Assume that the temperature of the gas remains constant.
4. At a temperature of 25°C and a pressure of 1 atm, a gas occupies a volume of 1.25 L. If the temperature in increased to 75°C and the pressure remains constant at 1 atm, what is the new volume that the gas occupies?
5. A bicycle pump has a barrel 65 cm long. On the upstroke, air is drawn into the pump at a pressure of 1 atm. On the downstroke, the air in the barrel is compressed. How long must the downstroke be to increase the pressure of the air in the barrel to 4 atm? Assume that the temperature of the air remains constant during the compression process.
6. Calculate the number of liters occupied at STP by 0.35 moles of nitrogen gas (N_2).
7. A 2.0-liter mixture of gases is produced from 2.0 liters of O_2 at 300 torr, 2.0 liters of N_2 at 500 torr, and 2.0 liters of Ar at 100 torr. What is the pressure of the mixture of gases in torr?
8. Two chambers are connected by a U-shaped manometer. The manometer is filled with oil having a density of 0.910 g/mL. The pressure of gas in Chamber A is 550 torr, and the oil in the arm of the manometer connected to Chamber A is 85 cm higher than the oil in the arm of the manometer connected to Chamber B. What is the pressure in Chamber B?
9. You have been assigned the task of decorating a ballroom with 250 helium-filled balloons. A 25-liter helium tank filled with helium at a pressure of 25.0 atm and three 100-balloon packages have been delivered to the ballroom. If each balloon, when filled, holds 2.4 liters of helium at a pressure of 1.09 atm, will you be able to fill all 250 balloons? Assume that the gas temperature stays constant as the balloons are filled.
10. From *Chemistry: Connections to Our Changing World* by H. Eugene LeMay, Jr., p. 430:

> The Scene: The local county fair. A magician has just announced to a crowd of people at the fair that he can tell one gas apart from another just by feeling the gases. "You can't feel a gas," came a voice from the crowd. "Well," said the magician, "for the price of my show, which happens to be one dollar, I'll demonstrate that I can identify a gas by the way it feels. If I can do it successfully, I get to keep the money. But if I fail, you will get a refund. How's that?"
>
> Their curiosity aroused, people quickly paid the money and within a very short time the magician's tent was full of eager spectators. They could not imagine how the magician would be able to tell two gases apart just by feel, without relying on color or odor. The magician stepped from behind a curtain holding two stoppered bottles. "Fill one bottle with xenon and the other with helium. Make sure you keep track of which gas is in which bottle."

A volunteer from the crowd filled the bottles from the compressed gas tanks that were conveniently secured to the stage. She identified each bottle with a secret mark and then gave the bottles to the magician.

The magician examined each bottle, took each stopper off briefly and stuck his finger into the bottles. "This is helium and this is xenon," he said. He was right! He performed the trick several times and never failed. How did he do it?

11. A recent advertisement appeared in a local newspaper about a local tire store filling car tires with pure nitrogen instead of atmospheric air. The ad claimed that tires filled with pure nitrogen leak slower than tires filled with atmospheric air. Is this claim valid? Give a supporting argument or reason for your answer.

LABORATORY EXERCISES

Magdeburg Hemispheres

Equipment Needed: Magdeburg hemispheres and a hand pump

Objective: To demonstrate the effects of a vacuum caused by a difference in air pressure inside and outside the Magdeburg hemispheres.

Laboratory procedure: Place both hemispheres together. Connect the pump to the valve on the hemispheres. Open the valve and pump down the interior of the hemispheres. Close the valve and disconnect the pump. Grasp both ends of the closed sphere by the handles and try to separate the hemispheres by applying force in opposite directions. What do you observe?

Now turn the valve handle to vent the interior of the sphere. The hemispheres should separate easily. Why does this happen?

Questions

1. Why is it so difficult to pull the hemispheres apart if the interior of the sphere is evacuated?
2. What is the theoretical pressure exerted by the atmosphere on the exterior of the Magdeburg hemispheres?
3. Why do the hemispheres come apart easily when the interior of the sphere is vented to atmosphere?

Boyle's Law Experiment

Equipment Needed: Boyle's law apparatus (a syringe and a pressure gauge) and graph paper.

Objective: To determine the relationship between pressure and volume at a constant temperature in a closed system.

Laboratory Procedure: First, you must determine the volume of the tubing and pressure gauge. You can accomplish this by setting the plunger at zero and noting the pressure reading. Now increase the volume by drawing back the plunger until the pressure drops to half its original reading. The volume has now doubled. The reading on the syringe will equal the volume in the tubing and gauge.

Second, set the plunger at its midpoint. Move the plunger forward to decrease the volume in equal volume increments (cubic centimeters) and record the pressure. Do the same in the other direction, increasing the volume.

Finally, plot the pressure versus volume (syringe reading plus tubing and gauge volume).

Question

1. What is the mathematical relationship between pressure and volume at a constant temperature?

Imploding Soda Can

Equipment Needed: Hot plate, plastic tub, water, ice, aluminum soda cans, and insulated glove (a potholder will suffice) or tongs to grasp the aluminum can.

Objective: To demonstrate the effect of an imbalance in air pressures inside and outside the soda can.

Laboratory Procedure: Make an ice water bath in the plastic tub. Heat the hot plate to a temperature that will boil water. Put a small amount of water into an empty aluminum can. Place the can on the hot plate and heat until steam escapes from the hole in the top of the can. Using the insulated glove or tongs, quickly invert the can and immerse it into the ice water bath. What do you observe?

Questions

1. What caused the can to implode?
2. Would the same effect be obtained if the can were not inverted when it was placed or dropped into the ice water bath?
3. What would have been the effect if the heated can had been inverted and immersed into a tub of *hot* water?

The Expanding Balloon

Equipment Needed: Rough vacuum system with a bell jar, and a small latex balloon.

Objective: Observe the effect of reduced ambient pressure on a partially inflated balloon.

Laboratory Procedure: Partially inflate the balloon and tie off the opening. Place the partially inflated balloon under the bell jar. Start the rough vacuum pump and pump down the bell jar. Observe what happens to the size of the balloon as the pressure in the bell jar is reduced.

Questions

1. Why did the balloon expand when the bell jar was evacuated?
2. Why did the balloon return to its original volume when the bell jar was vented?

Boiling Water at Room Temperature

Equipment Needed: Rough vacuum system with a bell jar, a beaker, water, and a thermometer.

Objective: Determine the effect of reduced ambient pressure on causing water to boil at room temperature.

Laboratory Procedure: Fill the beaker half full of water. Measure the temperature of the water in the beaker. Mark the height of the water on the side of the beaker. Place the half-filled beaker under the bell jar. Start the vacuum pump and pump down the bell jar. Observe what happens to the water in the beaker as the ambient pressure is reduced and base pressure is reached. Once you have made these observations, stop the vacuum pump and vent the bell jar to atmosphere. Remove the beaker of water and measure the temperature of the water.

Questions

1. Why did bubbles form in the water when the pressure in the bell jar was lowered?
2. What conditions had to exist in the bell jar in order for the water to boil?
3. Did you notice a volume change in the amount of water in the beaker?
4. Was there a temperature change in the water between the beginning of the experiment and the end of the experiment? How can you account for this?

CHAPTER 3

An Introduction to Vacuum Systems

3.1 INTRODUCTION

The purpose of this chapter is to set the stage for Chapters 4 and 5. As we study the rough and high-vacuum pressure regimes, we will approach each in the same manner. First, the gas load in the given pressure regime will be described. Then the different pumping mechanisms that can be used to remove gas molecules from the chamber and appropriate pressure gauges will be presented. A short discussion of valves and fittings will follow, and the chapter will close with system characterization and troubleshooting.

The chapter begins with a discussion of gas load. Contributions to the gas load in a system come from many different sources. In each pressure regime, certain gas sources play a bigger role than others. Section 3.2 provides an overview of vacuum pumps, and Section 3.3 does the same for vacuum gauges. More specific information on pumps and gauges will be presented in the following chapters. Section 3.4 surveys other vacuum system components needed to construct a complete vacuum system. Section 3.5 introduces two new terms used to describe vacuum systems, conductance and throughput. A simple

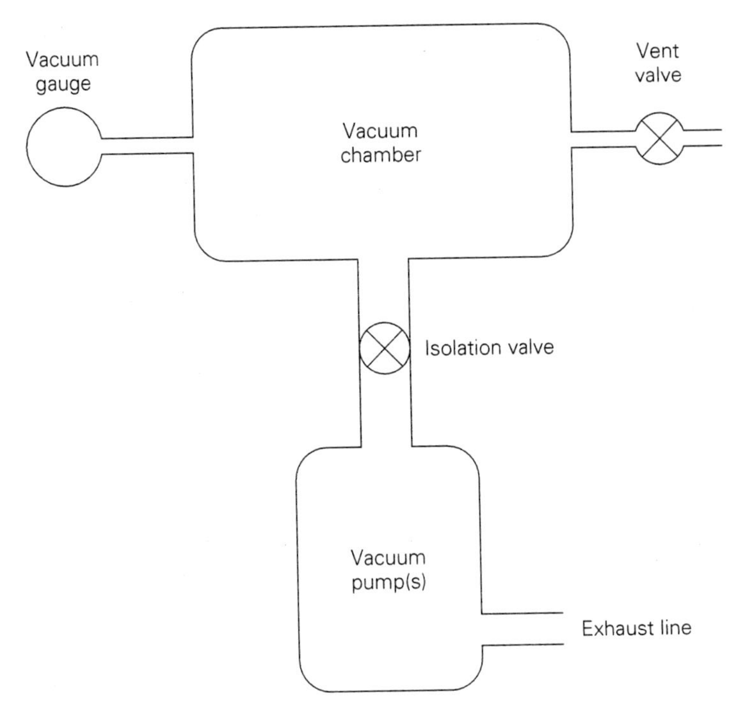

FIGURE 3.1
Simplified block diagram of a vacuum system.

vacuum training system is described in Section 3.6. And finally, Section 3.7 presents a rationale for using pump-down curves to characterize vacuum systems.

Vacuum systems come in many different designs and configurations. Figure 3.1 shows a simplified block diagram of a vacuum system. The chamber is connected to the vacuum pump(s) via a vacuum line constructed of some type of piping. Inserted into the vacuum line between the chamber and the pump is an isolation valve. The isolation valve allows the pump to run while the chamber is vented and allows the chamber to be under vacuum while the pump is not operating.

Two other vacuum components are attached to the chamber. One is a pressure gauge to measure the chamber pressure. The other is a vent valve that allows air to flow into the chamber so that the chamber pressure can be returned to atmospheric pressure, a necessary condition if the chamber is to be opened.

The simplified block diagram in Figure 3.1 provides a visual image of a vacuum system for the discussion in this chapter.

3.2 GAS LOADS

Vacuum pumps have to deal with a certain gas load in each pressure regime. The *gas load* is the amount of gas in the chamber that has to be evacuated. A number of sources contribute to the gas load in a vacuum system, and at any point in time, the total gas load is the sum of all the contributing sources. Let us consider some of the contributing gas sources (see Figure. 3.2).

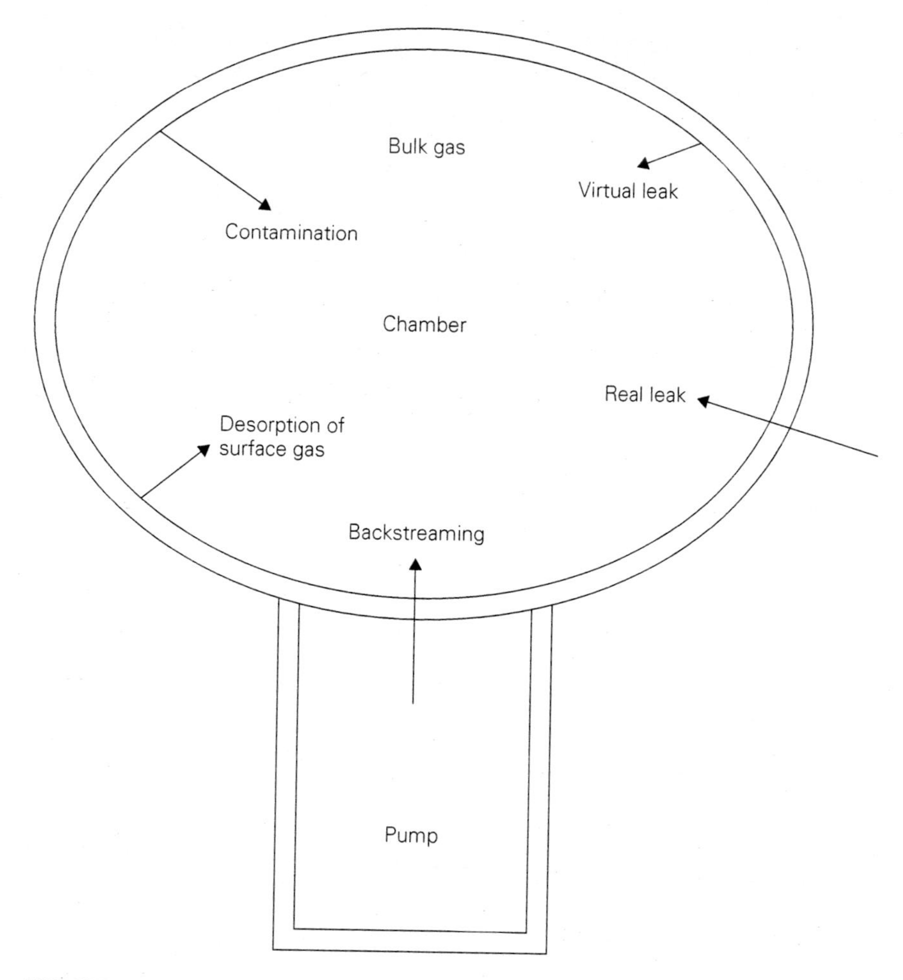

FIGURE 3.2
Sources that contribute to gas load in the rough vacuum regime.

Every system, if it has been opened to the atmosphere, begins at approximately 1 atm, or 760 torr. The gas load at this point is simply defined as the product of the pressure in the chamber times the volume of the chamber. Hence, the larger the chamber, the larger the corresponding gas load at a given pressure. Conversely, the smaller the chamber, the smaller the corresponding gas load.

The first component of gas load to be removed is the bulk gas. The *bulk gas* is the gas that is originally in the chamber. The amount of bulk gas is the product of the chamber volume times the pressure in the chamber. For example, if the chamber volume is 10 liters and the pressure is 760 torr, the amount of gas in the chamber is 10 liters × 760 torr, or 7600 torr-liters.

As the chamber is pumped down, the gas load decreases because the chamber pressure is dropping while the chamber volume remains constant. At the low end of the rough vacuum regime, the bulk gas has been reduced to a level where it is no longer the dominant contributor to gas load. At this point, water vapor desorbing from the chamber walls becomes the dominant source.

In the high-vacuum regime, other sources of gas come into play (see Figure 3.3). Outgassing and permeation can be significant contributors to gas load. *Outgassing* refers to the escape of gas that is trapped in solid materials. *Permeation* is gas that passes through materials used in the construction of vacuum systems, such as the elastomer O-ring seals in the flanges that connect the vacuum components.

Contributions to gas load due to leaks can be caused by virtual leaks or by real leaks. Leaks can occur in any of the vacuum regimes. We have all had experience with real leaks. *Real leaks* provide a path for air molecules to travel from outside the chamber to the inside of the chamber through a space in the wall or connection in a vacuum system. In contrast,

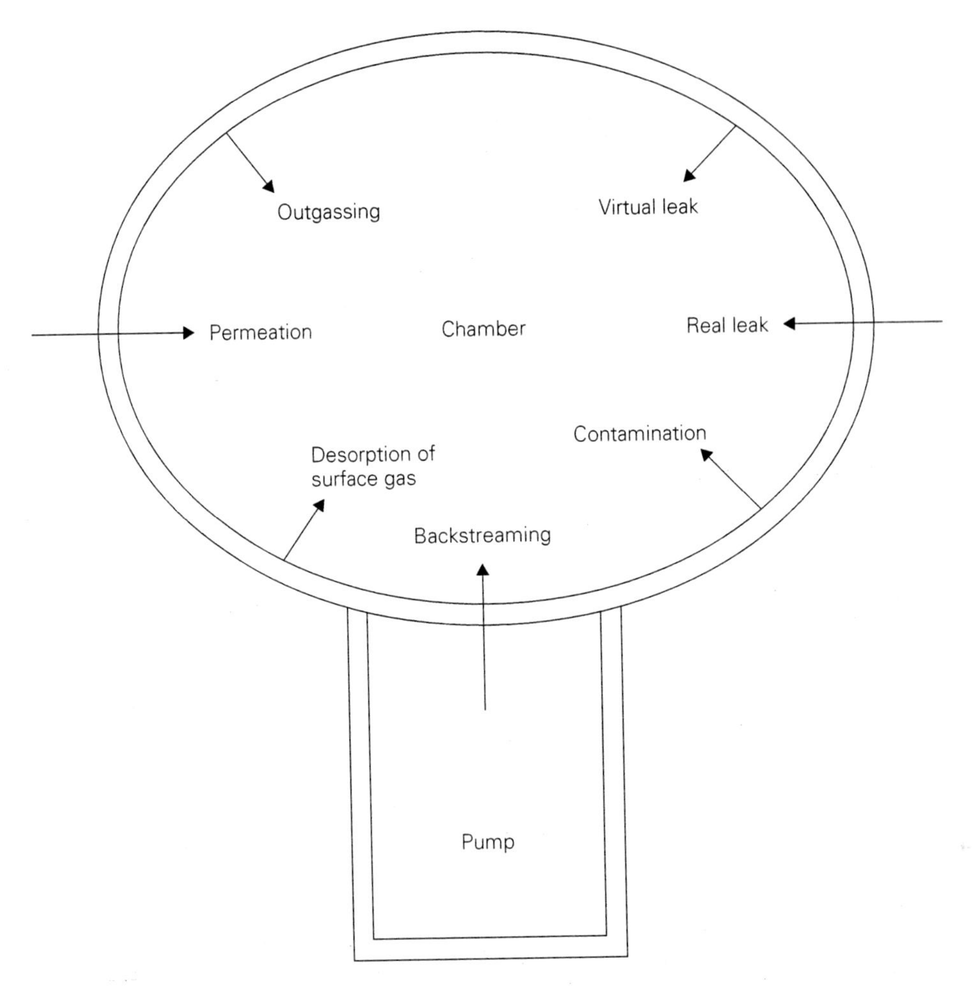

FIGURE 3.3
Contributors to gas load in the high-vacuum regime.

virtual leaks are not true leaks, but they act in a similar manner. *Virtual leaks* are pockets of trapped volumes of gas that escape slowly as pressure decreases in the chamber.

And finally, when a process is being run, the process chamber is pumped down to the desired pressure, and process gases are introduced into it at a specified flow rate. Some of the process gases react in the chamber, producing reaction by-products. Unreacted process gas molecules along with the reaction by-products add to the gas load seen by the vacuum system.

We will examine gas loads in more detail when we take a look at each pressure regime. Knowing the sources that contribute to gas load is important in operating, maintaining, and troubleshooting vacuum systems.

3.3 VACUUM CHAMBER

The *vacuum chamber* is usually the focal point in a process system. This is also true in vacuum system design. The walls of the chamber must be made thick enough to withstand the forces exerted by the atmosphere. In some cases, the chamber may be strengthened by adding additional supports, either internally or externally.

Chambers can be constructed of metal, glass, ceramic, or plastic components. Since all of these materials add to the total gas load in the chamber, selection is critical. Materials that do not bond well with water molecules and do not outgas very fast are ideal. This rules out plastic in high-vacuum applications. Metals such as stainless steel and aluminum have low outgassing rates in the range of 10^{-6} to 10^{-9} torr-L/sec · cm^2. Glass, or pyrex, also has a low outgassing rate in the range of 10^{-9} torr-L/sec · cm^2. For most applications in the high-vacuum regime, permeation through the chamber walls can be considered negligible. Another important consideration is the sealing material used to construct the chamber.

3.4 VACUUM PUMPS

Vacuum pumps come in a variety of types and styles, and pump selection depends heavily on the application in which the pump will be used. Figure 3.4 depicts a family tree of vacuum pumps. Although not a complete family tree, it shows the variety of vacuum pumps available today.

Vacuum pumps can be classified as either gas transfer pumps or entrapment pumps. *Gas transfer pumps* remove gas from the process chamber and exhaust it to an exhaust system in the building. That is, the gas passes through the pump, from inlet to outlet.

In contrast, the gas molecules taken in by an *entrapment pump* are "entrapped" within the pump. Entrapment involves changing the gas molecules in the gas phase to a solid by "freezing" them or chemically reacting them with a chemically reactive material such as titanium to form a solid compound that can stick to the interior surface of the pump.

Gas transfer pumps fall into two classifications: positive displacement vacuum pumps and kinetic vacuum pumps. *Positive displacement pumps* expand the volume of the chamber, seal off the volume of trapped gas, compress the trapped gas, and then exhaust the trapped gas. There are two types of positive displacement vacuum pumps: reciprocating and rotary. Diaphragm pumps and piston pumps are examples of reciprocating pumps, whereas rotary pumps include rotary vane pumps, rotary piston pumps, and Roots pumps.

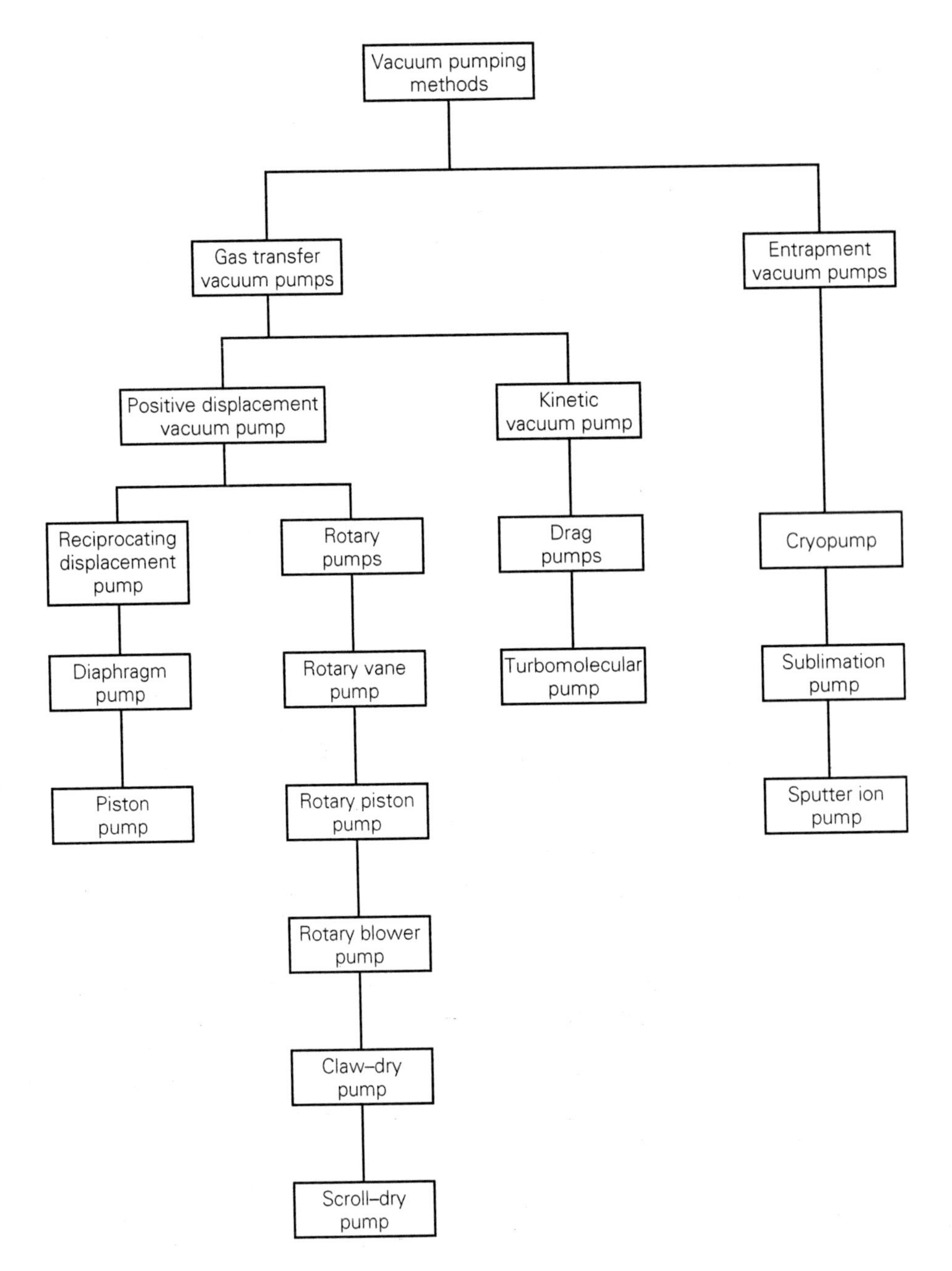

FIGURE 3.4
Family tree of vacuum pumps.

Kinetic vacuum pumps include turbomolecular pumps, or turbo pumps, for short. These pumps use high-speed turbine blades to move gas molecules through the pump. A roughing pump is then used to exhaust the gas molecules from the turbo pump to the exhaust system in the building.

A vacuum can also be created by condensing or freezing the gas. *Cryopumps* use this technique. Cryopumps use very cold surfaces to freeze the gas molecules, entrapping them on the inner surfaces of the pump.

A similar technique used to create a vacuum removes gas molecules from the gas phase through chemical reactions. For example, getter pumps use very reactive materials, such as titanium, to form solid materials from the gas molecules. These solid materials stick to the inner surfaces of the pump and effectively remove gas molecules from the gas phase in the process chamber.

Physical absorption is a technique used with cryogenic techniques. This technique uses highly porous materials, such as activated charcoal, activated alumina, and materials called zeolites, as gas sponges. The gas molecules travel into the porous structure of the material and are entrapped within the pump.

Given the myriad of vacuum pumps on the market today, selecting the "best" vacuum pump for a specific application can be a daunting task. Unfortunately, there will probably be no single pump that will satisfy all the process requirements. Here are some useful considerations when selecting a vacuum pump:

- What ultimate pressure must be reached during the process? Do the pumping speed versus pressure curves fit the process requirements?
- What is the anticipated gas load? Are there any problems associated with the gas load, for example, use of corrosive gases?
- Can the pump be installed in the intended space and with the desired orientation? Are there other considerations, such as vibration and noise, that might affect its installation?
- What are the pump's power requirements?
- Will the pump require a backing pump or other components such as valving?
- Can the pump be easily maintained? How often will the pump(s) require maintenance?
- What cleanliness level is required? Is an oil-free pump required?
- Will the process chamber be opened frequently?
- What will be the cost of ownership over the period of use, not just the original purchase price?
- What is the reliability and reputation of the vendor?

This is by no means a comprehensive list. Each application will have its own set of questions that have to be asked in order to address all the issues associated with your process. The questions in the list will give you a starting point in your selection process, or elimination process, as the case may be.

Data on the relative use of various types of vacuum pumps are difficult to obtain. However, *R&D Magazine* conducted a survey and published the results. Respondents to the survey used vacuum systems in the following applications: semiconductor devices (28.3% of the respondents), industrial coatings (27.5%), optical coatings (23.4 %), MEMS (20.8%), nanotechnology (20.8%), ceramics (16.4%), biomedical research (15.6%), lasers and electro-optics (15.2%), instrumentation (14.5%), plastics (13%), flat panel displays (12.6%), metallurgy (11.2%), and III-V–based semiconductor devices (10.8%). Obviously, since the percentages add up to more than 100%, some respondents were involved in more than one type of application.

In terms of operating pressure levels, fewer than 4% required operating pressures in the ultrahigh-vacuum regime (10^{-9} to 10^{-12} torr). About 23% operated in the lower portion of the high-vacuum regime (10^{-6} to 10^{-9} torr), and 28% operated in the upper portion of the high-vacuum regime (10^{-3} to 10^{-6} torr). The remaining 45% of the respondents operated in the rough vacuum regime (1 atm to 10^{-3} torr).

As a result of the wide range of operating pressure levels, the survey respondents used many types of pumps. Most commonly used pumps were cryogenic (45%), turbomolecular (45%), diffusion (44%), and rotary vane (35%). More rarely used pumps were Roots (25%), diaphragm (16%), ion (15%), dry compression (14%), getter (8%), scroll (7%), molecular drag (6%), and sorption (4%). Respondents cited ease of maintenance, corrosion resistance, and oil-free operation as the most important selection criteria. Particle tolerance, low vibration, and small footprint ranked low.

The results of the survey indicate that there are a wide variety of applications for vacuum pumps, and that a myriad of different pumps and pumping configurations are used in these applications. In selecting pumps, the application, project budget, and other considerations will determine which pump parameters and characteristics are most important. There is no "one-size-fits-all" solution, and it will take effort to find a suitable, if not optimal, solution to one's pumping needs.

3.5 VACUUM GAUGES

Measuring pressure in each regime requires the conversion of a parameter associated with the gas into an electrical signal that is then converted into a pressure readout. Vacuum pressure gauges differ in respect to operating pressure range, accuracy, and cost. All vacuum pressure measurement systems include a transducer and a gauge controller that performs the conversion process and provides the pressure readout. Figure 3.5 shows a family tree of common pressure gauges used on vacuum systems.

In the rough vacuum regime where the gas is relatively dense, two parameters are commonly used, the thermal conductivity of the gas and the force exerted by the gas on a surface. Examples of rough vacuum gauges that rely on the *thermal conductivity* of the gas are the thermocouple gauge, the Convectorr gauge, and the Pirani gauge. The *transduction* process involves the transfer of heat from a filament in the gauge as the gas molecules strike the filament. In a *thermocouple gauge,* this transfer of heat results in a temperature change in the filament that is sensed and converted to an electrical signal. In the case of a *Pirani gauge,* the change in temperature in the filament results in a resistance change, and this resistance change is converted to an electrical signal.

The *capacitance manometer* is a rough vacuum gauge that uses the force of the gas on a surface. There is a thin diaphragm inside the capacitance manometer. On one side of the diaphragm is a reference pressure, and on the other side is the gas in the chamber. A pressure differential causes the diaphragm to deform. The diaphragm forms one metal plate of a capacitor, and since $C = \varepsilon\frac{A}{d}$ where C is the capacitance, ε is the permitivity of

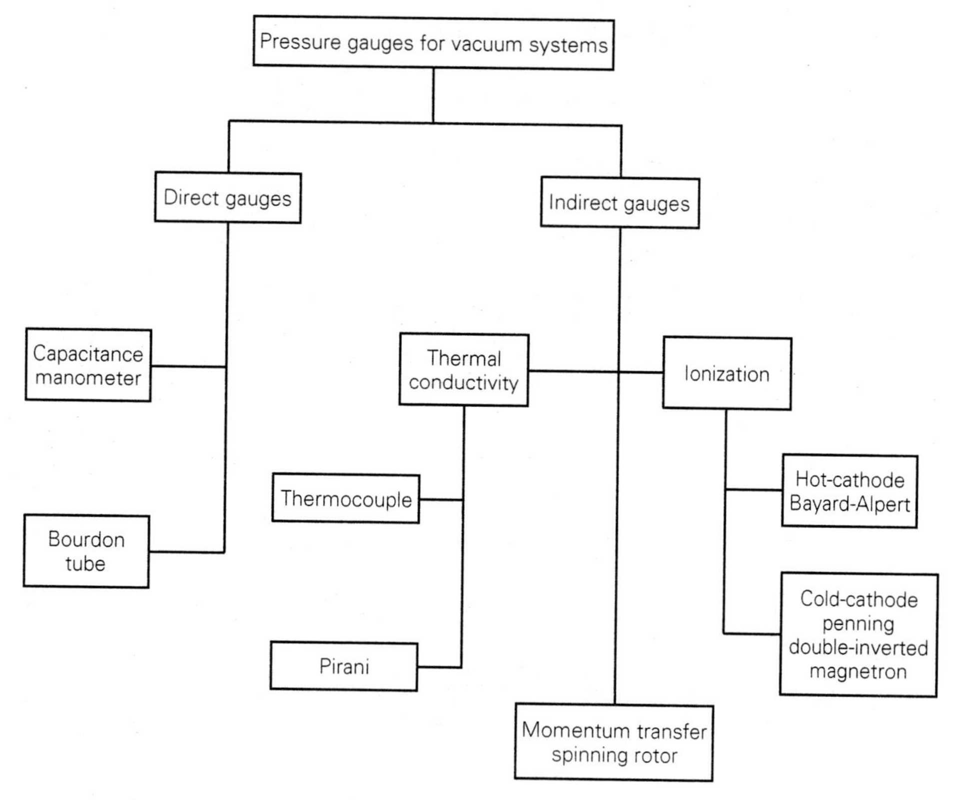

FIGURE 3.5
Family tree of pressure gauges.

the dielectric, A is the plate area, and d is the separation, any change in the separation caused by deformation of the diaphragm results in a change in the capacitance. The gauge controller converts this change in capacitance into an electrical signal and a pressure readout.

In the high-vacuum regime, the air is too rarefied to rely on thermal conductivity and force to measure pressure. High-vacuum gauges such as the Bayard-Alpert gauge sample the pressure by ionizing gas molecules within the gauge. At the high-pressure end of the operating range, a large number of ionizations occur and result in a large current in the collector terminal of the gauge. At the low-pressure end of the operating range, fewer ionizations occur, and the resulting current in the collector is smaller. The gauge controller converts the current in the collector to a pressure reading.

Unfortunately, measuring pressure is not as simple as connecting the transducer to the chamber and the transducer to the controller. Gases differ in their thermal conductivity and ease of ionization. As a result, the readout depends on the gas composition in the chamber. Correction tables for different gases are used to correct the pressure readings presented on the readouts of gauge controllers. We will look at examples in the next two chapters as we discuss pressure measurement in the rough and high vacuum pressure regimes.

Returning to *R&D Magazine*'s survey of vacuum applications, the use of vacuum gauges followed a traditional selection profile. More than half the respondents used thermocouple gauges, and slightly fewer than half used capacitance manometers. Only those respondents who operated in the high-vacuum regime had need of Bayard-Alpert and cold-cathode gauges. Other gauges used by the survey respondents included convection, UHV hot ionization, spinning rotor, inverted magnetrons, and a variety of mechanical gauges and mercury manometers.

3.6 VACUUM SYSTEM COMPONENTS

A vacuum system rarely consists only of a chamber, pump, and pressure gauge. Other components are usually added to configure the system to meet requirements and provide operating flexibility.

For example, a variety of valves provide vacuum systems with isolation, control, and venting capabilities. *Isolation valves* allow parts of the system to be isolated from other parts of the system. An isolation valve can be placed between the chamber and the pump so that the pumping on the chamber can be suspended without turning off the pump. Butterfly valves and throttling gate valves provide a way to control the flow of gas molecules and thus the pressure in the system. Air admittance valves provide a way to vent the chamber or admit gas into the system.

What should you consider when selecting a *valve?* Of course, there is the initial purchase price, but there are other considerations that affect the cost of ownership over the long haul. For example, the internal geometry of the valve, depending on its placement in the vacuum system, may add to the internal surface area of the chamber. Additional surface area means more water vapor in the chamber. Also, internal valve mechanisms are a source of particles and thus of contamination.

Another consideration when selecting the valve will be the material used to manufacture it. Valve bodies are made of brass, cast aluminum, machined aluminum, or stainless steel. Brass, because of its temperature limitations, porosity, and outgassing characteristics, is limited to the foreline of a vacuum system and to applications that do not use corrosive gases. Cast aluminum and stainless steel are used in a wide variety of pressure ranges. Stainless steel has the advantage of resisting many corrosive gases.

Three common types of valves are gate, poppet, and butterfly. *Gate valves,* configured with either circular or rectangular openings, provide the highest conductance. They feature a straight-through aperture. *Poppet valves,* also called block valves, have a sealing component that is lowered into the cylindrical body to seal against a seat inside the valve. Because of their simple design, poppet valves have a long cycle life. Poppet valves have a lower conductance than gate valves because they force the gas to make turns as it moves through the valve. *Butterfly valves,* along with throttling gate valves, are variable conductance valves that provide a means of controlling pressure in a system.

Other vacuum system components provide the plumbing that connects other parts. These include pipeline components, such as nipples, elbows, tees, crosses, couplings, reducers, and adapters, and come in industry-standard sizes.

The connections between these components also come in a variety of styles. The most common are the conflat flange (CF), the Klamp Flange (KF) and the ISO flange (see

FIGURE 3.6
Vacuum flange fittings.
Source: BOC Edwards Catalog, p. 446.

Figure 3.6). The conflat flange uses a copper gasket to form a seal, and the components are bolted together. The KF flange, on the other hand, uses an elastomer O-ring to form the seal, and a KF clamp is used to hold the components together and compress the elastomer material. ISO flanges are used for tubing requirements that generally exceed 2 in. in outside diameter, are similar to KF flanges that use a centering ring with an elastomer O-ring, and use a set of single- or double-claw clamps to hold the components together.

3.7 CONDUCTANCE

Pumping down a vacuum system means moving gas from the chamber through the pumping system. The term *conductance* is used to describe this ability of the pump and piping to move a given volume of gas in a given unit of time, and is expressed in a unit of volume per unit of time—for instance, liters per second (L/sec), cubic feet per minute (cf/m), or cubic meters per hour (m^3/hr).

The conductance of a pipe depends on its geometry (for example, length, diameter, and whether the pipe is straight or has bends). The conductance also depends on the operating pressure regime (such as rough vacuum or high vacuum) and whether the gas flow is viscous or molecular. These two types of gas flow will be described later in this book.

Consider the simple vacuum system shown in Figure 3.7. Suppose the piping between the chamber and the pump is capable of transporting gas at 400 L/sec when the isolation valve is fully open. The vacuum pump used in the system has a pumping speed of

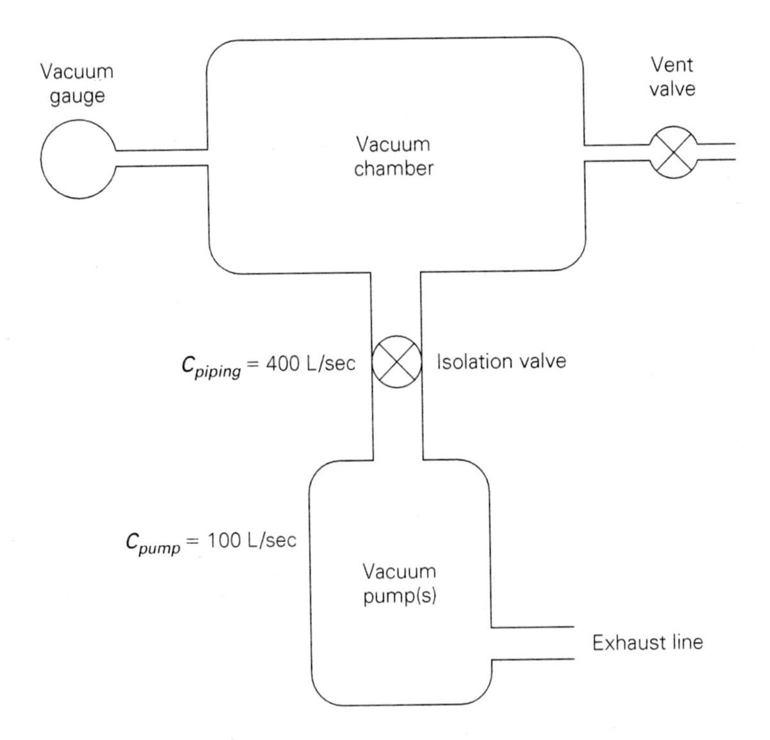

FIGURE 3.7
Conductances in a simple vacuum system.

100 L/sec. Since both of these properties deal with the ability to move gas from the chamber, we can say that the conductance of the piping (C_{piping}) is 400 L/sec and the conductance of the pump (C_{pump}) is 100 L/sec.

The two conductances, C_{piping} *and* C_{pump}, are in series in our system. From our study of electric circuits, we recall that conductances in series can be combined in a manner similar to resistors in parallel. That is

$$\frac{1}{C_{eq}} = \frac{1}{C_1} + \frac{1}{C_2} + \cdots + \frac{1}{C_n}. \qquad \textbf{(3.1)}$$

where n is the number of parallel elements.

Returning to our simple pumping system example, the net conductance of the system can be found by combining the two series conductance as follows

$$\frac{1}{C_{net}} = \frac{1}{C_{piping}} + \frac{1}{C_{pump}} \qquad \textbf{(3.2)}$$

where C_{piping} is the conductance of the piping, and C_{pump} is the conductance of the pump.

Substituting our known values for the two conductances yields

$$\frac{1}{C_{net}} = \frac{1}{400\ \text{L/sec}} + \frac{1}{100\ \text{L/sec}}$$

Solving for C_{net},

$$C_{net} = \frac{(400\ \text{L/sec})(100\ \text{L/sec})}{400\ \text{L/sec} + 100\ \text{L/sec}}$$

$$C_{net} = 80\ \text{L/sec}$$

C_{net} is the volumetric pumping capacity of the system. Unfortunately, it does not tell us the quantity of gas flow. For this, we use another term called *throughput.* Throughput, Q, is defined as

$$Q = P \times C_{net} \qquad \textbf{(3.3)}$$

where P is the pressure.

Since C_{net} has the units of volume per unit time, V/t, throughput, Q, is equivalent to

$$Q = \frac{PV}{t}. \qquad \textbf{(3.4)}$$

Common units for throughput, Q, are torr-liters per second, Pa-liters per second, or atm-ft^3 per minute.

Note that if the volume is constant and the same unit of time is used, then $Q \propto P$. This means that throughput will decrease as pressure decreases. In other words, it is getting harder to remove gas from the chamber because as pressure decreases, there are fewer and fewer gas molecules per unit volume.

We will return to the topic of conductance in the chapters that follow. For each pressure regime, rough and high vacuum, conductances will be calculated using mathematical formula specific to each regime. For now, let us move on to describe a simple vacuum system that is used for vacuum system training at community colleges and universities across the country.

3.8 A SIMPLE VACUUM SYSTEM

We will now discuss a simple vacuum system used in college teaching laboratories. The vacuum training system shown in Figure 3.8 is manufactured by Varian Vacuum Products.

The chamber is a quartz bell jar resting on a stainless-steel baseplate. Enclosing the bell jar is a metal enclosure that serves as a protective barrier should the bell jar crack.

The vacuum trainer has two pumps: an MDP 30 diaphragm pump and a Turbo V70LP turbomolecular pump. The MDP 30 is a four-stage diaphragm pump with a nominal pumping speed of 30 L/min and an ultimate pressure of approximately 0.5 torr. The MDP 30 is connected to the chamber through a separate roughing line.

The TurboV70LP is a small turbomolecular pump that operates at a rotational speed of 75,000 rpm and a pumping speed for nitrogen of 60 L/sec. Together, the Turbo V70LP and the MDP 30 are capable of pumping down the bell jar into the 10^{-6} torr pressure range.

Rough vacuum pressure measurements are made using thermocouple gauges. One is placed in the roughing line, and another is situated below the baseplate to measure chamber pressure. High-vacuum pressure measurements are made with a hot-cathode Bayard-Alpert gauge. The gauge controller can be either a SenTorr Vacuum Gauge Controller or a Multi-Gauge Vacuum Gauge Controller.

An optional addition to the vacuum trainer is a residual gas analyzer, such as the Stanford Research Systems' 100 amu RGA. The RGA measures the partial pressure of the gases in the bell jar and provides a graphical output of partial pressures versus mass-to-charge ratio.

The trainer incorporates two different types of fittings: KF fittings in the roughing line and Conflat fittings in the high-vacuum line.

Five valves, four manually operated and one automatic, are included in the system. One block valve provides isolation in the roughing line to the chamber, and another provides isolation between the diaphragm pump and the turbomolecular pump. A high-vacuum gate valve provides isolation between the turbomolecular pump and the chamber. A vent valve allows atmospheric air to enter the chamber. An automatic vent valve is connected to the turbomolecular pump. Finally, a gas inlet valve provides a means of supplying a gas to the bell jar.

MKS Instruments also manufactures and markets a vacuum training system. The MKS Vacuum Training System, Type VTS-1B, shown in Figure 3.9, is designed as a medium vacuum process system with provisions for either manual or automatic pressure control. The VTS-1B can be modified or expanded by adding the High Vacuum Modification Package and a Vac-Check Quadrupole RGA. Once these modifications are added, the VTS-1B is comparable in performance to the Varian Vacuum Trainer. The major difference between the two systems is whether a component-type system is more suitable for the intended application. More information on the VTS-1B can be found at the MKS Web site, www.mksinst.com.

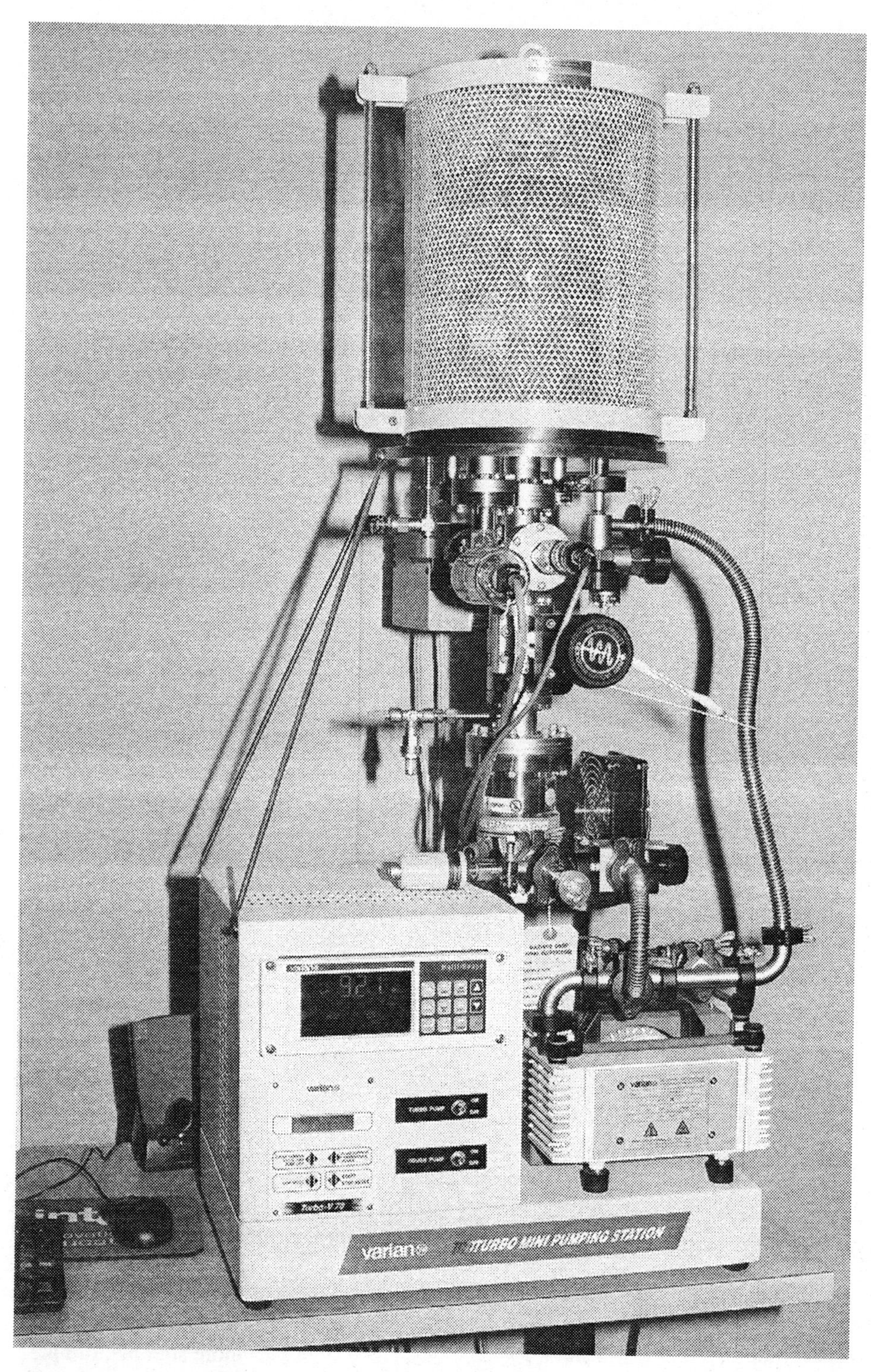

FIGURE 3.8
Varian Vacuum Training System.

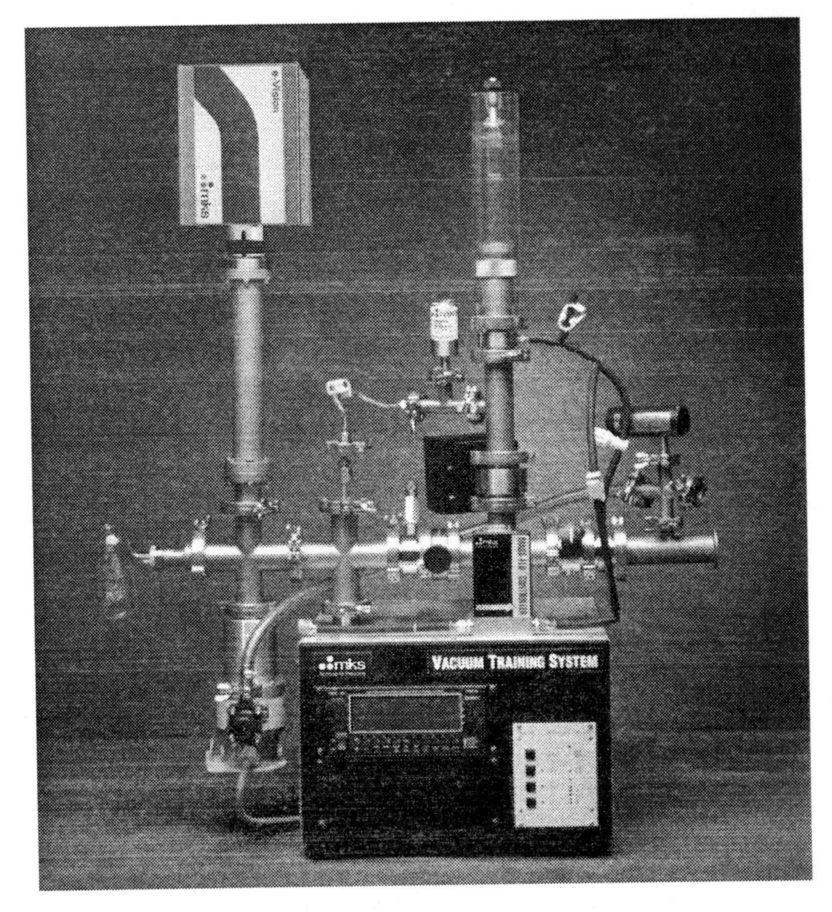

FIGURE 3.9
MKS VTS-1B Vacuum Training System. Used by permission, MKS Instruments.

3.9 CHARACTERIZING VACUUM SYSTEMS

Knowing your vacuum system is the first step in detecting and addressing problems that occur in vacuum systems. Qualitatively, this can be as simple as hearing unusual sounds from your system, just as you might hear odd noises in your car when you start or drive it. Unfortunately, problems with vacuum systems cannot always be detected by listening.

Another method is to keep records of vacuum system performance. For example, a system log might include the time-to-pressure or pump-down times recorded at optimum performance. This offers simple tracking data that will indicate when system performance has changed, such as when the system seems to pumping down slower than usual or is not pumping down quite as far as it used to.

Phil Danielson, in an October 1998 article entitled "The Value of Pumpdown Curves," describes the industry practice of using "pump-down curves" to characterize vacuum systems and diagnose problems. A *pump-down curve* is a graph of pressure versus time that describes the performance of a vacuum as the system is pumped down from initial pressure to base pressure. Using a pump-down curve taken when the system is operating optimally, future pump-down curves can be compared to the optimum curve to monitor system performance. If the two curves are identical or very similar, chances are that the system is operating normally and there is no problem. On the other hand, if the two curves differ, this can indicate that a problem has occurred in the system and corrective action should be considered.

As an illustration, Danielson describes the following scenario. Consider a high-vacuum system consisting of an oil-sealed mechanical pump backing a high-vacuum pump (for example, a turbomolecular pump). The time-to-pressure measurements have been fairly consistent, and the system seems to be operating normally without problems. However, the shape of a pump-down curve taken recently indicates that the upper part of the curve (the roughing portion) might be slowing. The question is, does this indicate a problem in the system?

The problem is described by Phil Danielson as follows:

> It might well be. Let's say that the humidity has slowly increased and the rough pump pumping efficiency is falling off due to a slow buildup of condensed water vapor in the roughing pump. If it is an oil-sealed mechanical pump, it is probably time to perform a gas ballast to remove the condensed water before the high vacuum pump's gas handling ability is overwhelmed. Let's also say a turbo-pump is being used and it is still able to handle the water load and produce an acceptable time-to-ultimate performance. If the water load increases just a little bit more, performance will start to drop off. Murphy's law says that will happen right in the middle of your process. This could have been avoided by acting upon the early warning provided by a simple and routine (pumpdown) curve comparison.

Several other scenarios described in the article also point to the utility of pump-down curves. The message is not that constant comparison of curves will solve all vacuum system problems, but rather to show the advantages of their use. The small amount of time and effort that goes into checking pump-down curves can easily be offset by not having to solve a problem after the fact or tracking the wrong suspected problem. A good reference pump-down curve is also valuable following any change in the system, such as cleaning, or replacing a system component. The "good" pump-down curve can serve as a benchmark to indicate that the system has been restored to its optimal operating condition.

SUMMARY

This chapter described the general components used to construct vacuum systems: vacuum chamber, vacuum pumps, pressure gauges, and valves and fittings. The amount of gas from all sources to be removed from the system was defined as the gas load. These sources of gas include the bulk gas, desorption of surface gas, outgassing, permeation, contamination, leaks, and backstreaming.

Vacuum pumps are used to remove gas from the chamber. Vacuum pumps come in a variety of designs and work over a limited range of gas pressures. Rough vacuum pumps are typically compression-type pumps that trap a volume of gas, compress it, and then exhaust it from the system. High-vacuum pumps are either kinetic-type pumps (such as turbomolecular pumps) or capture pumps (such as cryogenic or cryopumps).

Vacuum pressure gauges also come in a variety of designs. In the rough vacuum regime, gauges operate either through the force exerted by a gas on a solid wall or by sensing the loss of energy from a filament via thermal conduction by the gas molecules. In the high-vacuum regime, pressure gauges operate by ionizing the gas molecules.

Parts that connect the chamber to the pumps and gauges also come in a myriad of designs. Fittings include KF, conflat, and ISO designs, and piping in the form of tubes, elbows, tees, and other forms. The valves used in the rough vacuum regime are typically block valves, whereas high-vacuum valves may be of the liner gate variety.

The terms "conductance" and "throughput" were introduced to describe the flow of gas from the chamber. Conductance is a volumetric measure given in units of volume per unit time. Throughput describes the amount of gas moving through the pumping system and is measured in units such as torr-liters per second.

The chapter concluded with a discussion of the use of pump-down curves to characterize vacuum systems. The health and well-being of a vacuum system can be assessed by comparing pump-down curves from different points in time.

Now that we have a general sense of what makes up a vacuum system, let us look at vacuum systems for the rough vacuum pressure regime and the high-vacuum pressure regime. We will begin with the rough vacuum pressure regime in chapter 4.

BIBLIOGRAPHY

Comello, Vic. "Taking Vacuum Valves Seriously." *R&D Magazine,* March 1998, p. 61.

Danielson, Phil. "Gas Loads in Vacuum Systems." *Vacuum & Thin Film,* October 1998, pp. 36–39.

———. "The Value of Pumpdown Curves." *Vacuum & Thin Film,* October 1998, pp. 12–14.

———. "Matching Vacuum Pump to Process." *R&D Magazine,* November 2001, pp. 53–55.

O'Hanlon, John. *Basic Vacuum Practice,* 3rd ed, Varian Associates, Palo Alto, CA, 1992.

———. "Advanced Vacuum Practices." *Course Notes.* Rev. 6, December 3, 1996. Varian Vacuum Products, Lexington, MA.

———. *Product Catalog 2003/04.* Publication No. C100-03-895. BOC Edwards, 2003.

———. *A User's Guide to Vacuum Technology,* 3rd ed. Hoboken, NJ, Wiley-Interscience, 2003.

Studt, Tim. "Exclusive Survey Reveals Move to High-Tech Solutions," *R&D Magazine,* www.rdmag.com.

PROBLEMS

1. When referring to a vacuum system,
   a. Define the term "gas load."
   b. In the rough vacuum regime, identify the major contributors to gas load.
   c. In the high-vacuum regime, identify the major contributors to gas load.
2. What is the net conductance of a simple vacuum system in which the conductance of the piping is 300 L/sec and the pumping speed of the pump is 600 L/sec?
3. What would be the effect of changing the pump in Problem 2 from a 600-L/sec pump to a much larger pump having a pumping speed of 1,800 L/sec? What percentage improvement can be achieved?
4. Pump-down curves are used as a diagnostic tool in maintaing vacuum systems.
   a. What information do pump-down curves provide?
   b. Give an example of a problem that could be diagnosed using a pump-down curve?

CHAPTER 4

Rough Vacuum Regime

4.1 INTRODUCTION

Any vacuum system that has been opened to the atmosphere will have a chamber pressure of approximately 1 atm, or 760 torr. When pumped down, the first pressure regime the chamber pressure will pass through is the *rough vacuum* pressure regime. For our discussion, we will define the rough vacuum regime as pressures ranging from 759 torr to 1 millitorr, about six orders of magnitude for pressure. Over this pressure range, the molecular density of gas molecules

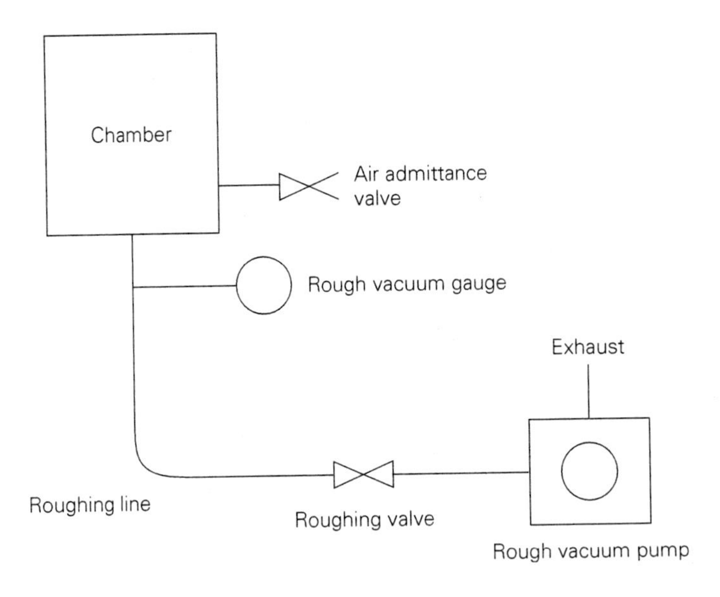

FIGURE 4.1
Rough vacuum system.

changes from a high of 3×10^{19} molecules per cubic centimeter (cm^3) at 759 torr to a low of 4×10^{13} molecules per cm^3 at 1 millitorr. Over the same pressure range, the mean free path of the gas molecules increases from 2.5×10^{-5} mm at 759 torr to 5.1 cm at 1 millitorr.

The components of a rough vacuum system are shown in Figure 4.1. The mechanical *roughing pump* is capable of reducing the chamber gas beginning pressure of 760 torr down to an ultimate pressure specified by the pump manufacturer. A nominal pumping speed in either liters per minute or cubic feet per minute will indicate the rate at which the bulk gas can be evacuated from the chamber. The inlet of the roughing pump is connected to the chamber by tubing or piping. An *isolation valve* is typically placed in the roughing line to provide means of isolating the pump from the chamber by blocking the pathway for the gas to flow between the chamber and the pump. The outlet of the roughing pump is usually connected to the *house exhaust system,* or if the chamber contains only room air, the exhaust can be left open to the room.

Two other features are needed to complete a rough vacuum system. First, there must be a way of measuring the chamber pressure. This is accomplished by connecting a rough vacuum gauge and gauge controller to the chamber. The *rough vacuum gauge* measures the chamber pressure and sends an electrical signal to the gauge controller. The *gauge controller* converts the electrical signal to a pressure reading and displays it.

Second, there must be a way of returning the chamber to atmospheric pressure so that the chamber can be opened after a chamber pump down. To provide this capability, a *vent valve* is connected directly to the chamber. When the vent valve is opened, air can flow in and return the chamber to room pressure.

In the following sections, we will examine each component of our rough vacuum system. Then we will look at rough vacuum system operation and conclude with a short section on troubleshooting rough vacuum systems.

4.2 GAS LOAD IN THE ROUGH VACUUM REGIME

The major components of gas load in the rough vacuum regime are the bulk gas, desorption of surface gas, and leaks (see Fig. 4.2). For the time being, we will assume that the integrity of the vacuum system has not been compromised and there are no leaks. That leaves the bulk gas and desorption of surface gas, mainly water in most cases, as the major contributors to our gas load. Other contributors to gas load, namely outgassing, diffusion, permeation, and backstreaming, are considered to make negligible contributions to gas load in the rough vacuum regime.

The bulk gas can be determined by multiplying the volume of the chamber (plus volumes open to the chamber) times the chamber pressure. The amount of gas is measured in

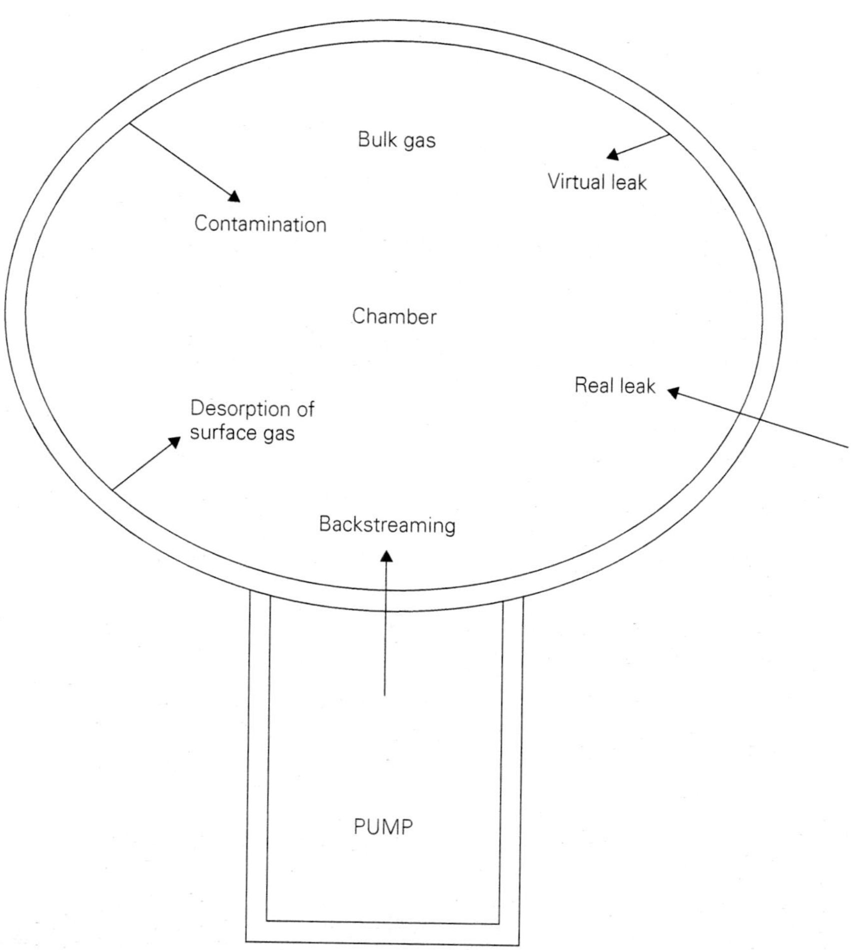

FIGURE 4.2
Gas load in the rough vacuum regime.

torr-liter units, or an alternative volume-pressure product. For example, a 5-liter chamber at 1 atm (760 torr) has 5 liters × 760 torr, or 3,800 torr-liters of gas.

The amount of surface gas is harder to quantify, for it depends on the surface area, the humidity of the air allowed into the chamber, and the time the chamber was vented to the atmosphere. We may have little control over these variables. Nevertheless, if the system has been vented to room air, water vapor from the air has coated the inner surfaces of the chamber. Many monolayers of water are on the interior surfaces of the chamber, piping, and other system components. As the chamber is pumped down, the water on the inner surfaces of the chamber will begin to desorb from the surface and enter the gas phase. Once in the gas phase, it can be pumped out of the chamber. Heating or baking the system aids desorption of water from the inner surfaces of the chamber.

4.3 PUMPING DOWN A ROUGH VACUUM SYSTEM

To pump down the vacuum chamber, the following procedure can be used. Beginning with the chamber at atmospheric pressure, the first step is to close the vent valve, if it isn't already closed. The rough vacuum gauge should be calibrated to 760 torr. The roughing valve should also be closed. In this configuration, the chamber is isolated from the atmosphere and the vacuum pump, and the vacuum gauge should indicate a pressure of 1 atm, or 760 torr, in the chamber.

Prior to beginning the pump-down process, the rough vacuum pump is started and the roughing line up to the rough vacuum valve is pumped down. If the volume of the line up to the valve is small, the pump-down time should be very short. To begin the chamber pump down, the roughing valve is opened. Once the roughing valve is opened, the vacuum pump will begin removing gas molecules from the chamber, and the pressure gauge will indicate a decreasing pressure in the chamber. The pressure in a small chamber will drop quickly and then gradually plateau at the base pressure of the system. The pump-down time for a system will depend on the size of the gas load and the net pumping speed. Figure 4.3 shows the general shape of the pressure versus time for a rough vacuum system.

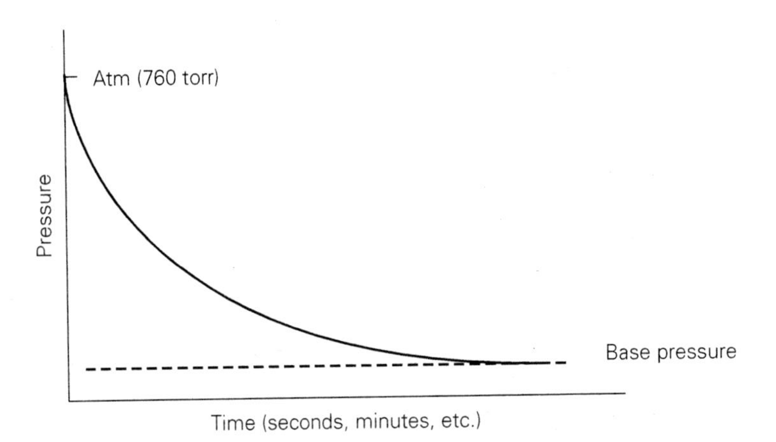

FIGURE 4.3
Rough vacuum pump-down curve.

4.4 ROUGH VACUUM PUMPS

Rough vacuum pumps are called *positive displacement* vacuum pumps. They operate by expanding the volume of the chamber by the volume within the pump, isolating the expanded volume within the pump, compressing the trapped gas, and then removing the trapped gas through the outlet of the pump.

To illustrate this process, consider a piston-type pump connected to a 10-liter chamber. The volume within the piston pump is 1 liter. Figure 4.4 shows the pump-chamber arrangement. What will be the pressure in the chamber after five strokes of the piston pump?

Initially, the chamber pressure is 1 atm, or 760 torr. The piston pump piston is pushed in so that the interior volume of the pump is zero. The pressure gauge reads a pressure of 760 torr.

We will now examine what happens in the vacuum system, one stoke of the piston pump at a time:

Stroke 1. When the Piston in the pump is drawn back, the 1-liter volume of the pump is added to the 10-liter volume of the chamber, making the total volume equal to 11 liters. The gas, originally enclosed in 10 liters, now fills 11 liters. The pressure in the chamber decreases to

$$P(\text{after 1 stroke}) = \frac{10}{11}(760\ \text{torr})$$

$$P(\text{after 1 stroke}) = 690.9\ \text{torr}$$

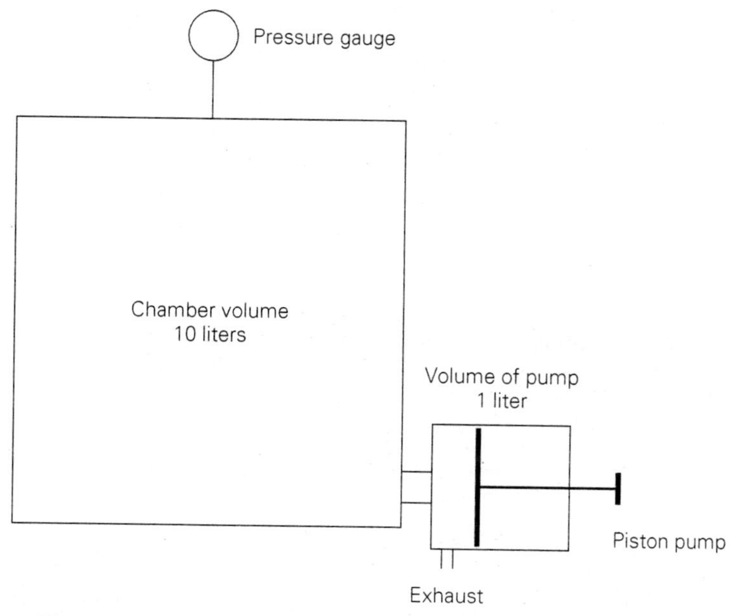

FIGURE 4.4
Piston pump connected to a 10-liter chamber.

The liter of gas in the pump is isolated, and then, as the piston is pushed forward and returned to its original position, the trapped gas in the pump is compressed and exhausted through the outlet valve of the pump.

Stroke 2. When the piston is drawn back again, the volume of the gas is again expanded from 10 liters to 11 liters. The pressure of the gas again decreases. The new pressure is

$$P(\text{after 2 strokes}) = \frac{10}{11}(690.9 \text{ torr})$$

$$P(\text{after 2 strokes}) = 628.1 \text{ torr}$$

The liter of gas in the pump is again isolated, and then, as the piston is pushed forward and returned to its original position, the trapped gas in the pump is compressed and exhausted.

Stroke 3. Again, when the piston is drawn back, the gas volume expands, and gas pressure decreases. The new pressure is

$$P(\text{after 3 strokes}) = \frac{10}{11}(628.1 \text{ torr})$$

$$P(\text{after 3 strokes}) = 571.0 \text{ torr}$$

Again the liter of gas in the pump is isolated, the piston is returned to its original position, and the trapped gas in the pump is compressed and exhausted.

Stroke 4. Once again, the gas volume expands. The pressure decreases to

$$P(\text{after 4 strokes}) = \frac{10}{11}(571.0 \text{ torr})$$

$$P(\text{after 4 strokes}) = 519.1 \text{ torr}$$

Then, gas isolation, compression, and exhaust are repeated.

Stroke 5. The expansion stroke repeats, and the pressure decreases to

$$P(\text{after 5 strokes}) = \frac{10}{11}(519.1 \text{ torr})$$

$$P(\text{after 5 strokes}) = 471.9 \text{ torr}$$

Finally, the liter of gas in the pump is again isolated, compressed, and exhausted.

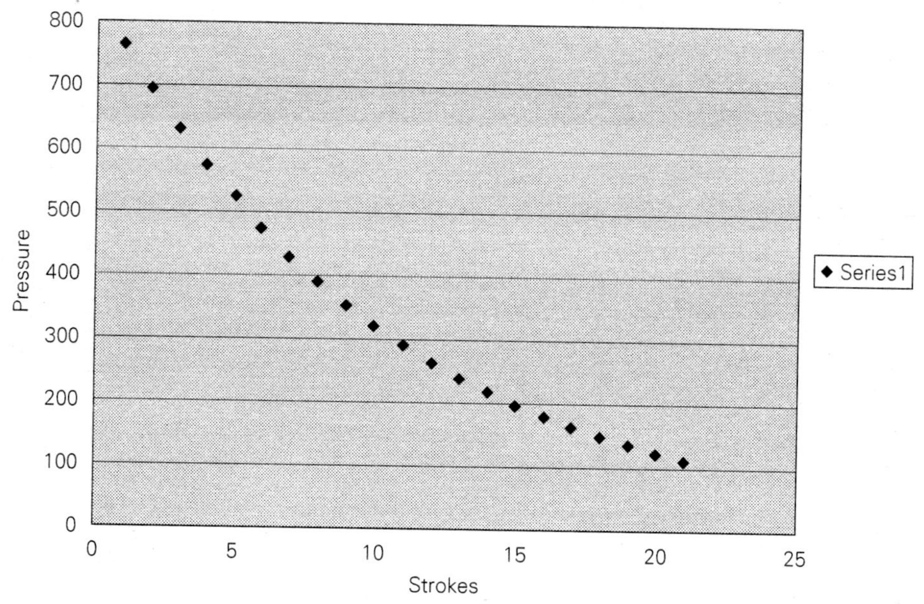

FIGURE 4.5
Sample piston pump pump-down curve.

Hence, the pressure in the chamber after five strokes of the piston pump is approximately 472 torr. If this example is continued for twenty strokes and a pressure versus strokes graph is plotted using a semi-log format, the exponential decay in pressure will be seen. The curve, as shown in Figure 4.5, will asymptotically approach some yet-to-be-determined base pressure.

Positive displacement vacuum pumps all use this expansion, isolation, compression, and exhaust cycle to move the gas from the inlet to the outlet of the vacuum pump. Let us examine the pumping action of four types of rough vacuum pumps: rotary vane, Roots blower, scroll, and diaphragm.

4.4.1 Rotary Vane Pump

An oil-sealed rotary vane pump is shown in Figure 4.6. The pumping mechanism in a rotary vane pump is enclosed by a stator. Inside the stator is an offset rotor that is mechanically coupled to a drive motor. The drive motor turns the offset rotor and spring-loaded vanes. A small amount of oil seals the space between the end of the sliding vane and the inner surface of the stator. As the rotor turns, a sealed space within the pump is created.

FIGURE 4.6
Rotary vane pump.
Source: (Varian Vacuum Technologies, *Product Catalog 2000*, p. 24. SD-700 Mechanical Pump)

Let us examine the operation of a rotary vane pump in more detail. A cross-sectional view of a rotary vane pump is shown in Figure 4.7(a). As the sliding vane passes by the opening to the inlet port, gas molecules begin to flow into the space behind the sliding vane. As the space behind the sliding vane increases, more gas molecules flow into this space. Eventually, the other end of the vane passes the inlet opening and seals off the volume of gas molecules that have moved into the pump.

As the rotor continues to rotate, the trapped volume of gas is gradually compressed and the pressure of the trapped gas increases. The compressed gas is then opened to the exhaust port. If the pressure of the trapped gas is greater than the pressure on the other side of the discharge valve, the discharge valve opens and the trapped gas escapes into the exhaust line. However, if the pressure of the trapped gas is insufficient to open the gas discharge valve, then the gas stays in the pump and no gas is exhausted from the pump. Figure 4.7(b) shows snapshots in time illustrating the movement of gas through the pump. First, gas enters the pump from the inlet port. Then the trapped gas is sealed off and compressed as the volume it occupies shrinks. Finally, the compressed gas exits the pump through the exhaust port. This process is repeated twice for each rotation of the sliding vane.

If the discharge valve does not open, there is a risk of vapor condensation and contamination of the pump oil. If the oil becomes contaminated with water or other impurities, these contaminants will increase the vapor pressure and make it impossible, or at least more difficult, to reach satisfactory base pressures.

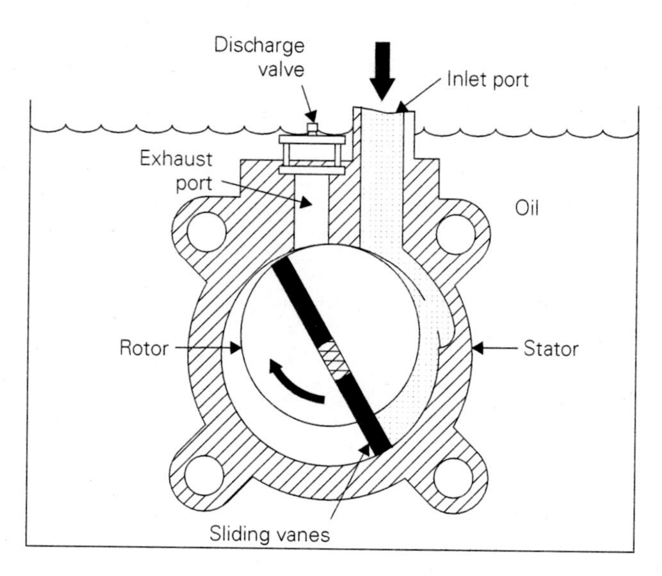

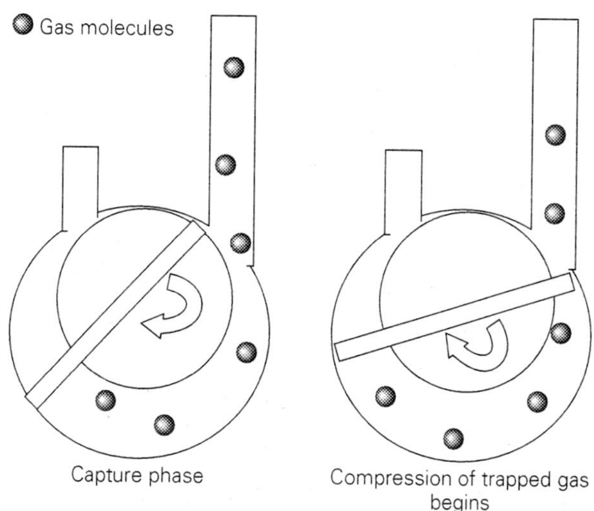

FIGURE 4.7(a)
Cross-section and pumping sequence of rotary vane pump.
Source: (Basic Vacuum Practice, 3rd Edition, Varian Vacuum Products.)

A gas ballast feature is used to ensure that the discharge valve opens during each and every cycle. The ballast valve is opened during each cycle, allowing a quantity of air or other gas to be admitted during the compression cycle. This extra gas ensures that the pressure of the compressed gas is great enough to open the discharge valve, allowing condensable vapors to exit the pump before they condense inside the pump.

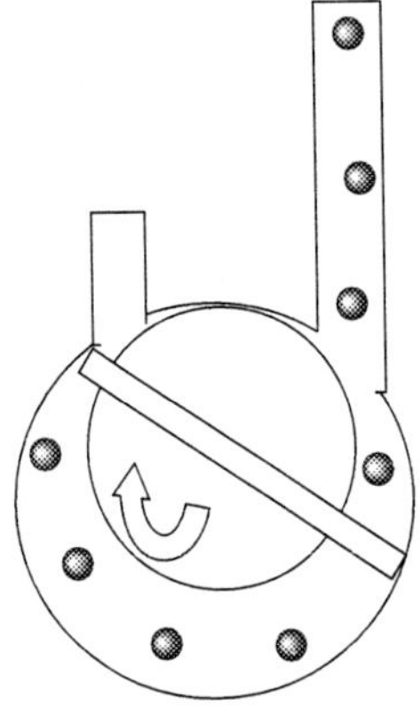

Compression of trapped gas continues, new gas molecules enters the pump behind rotating vane

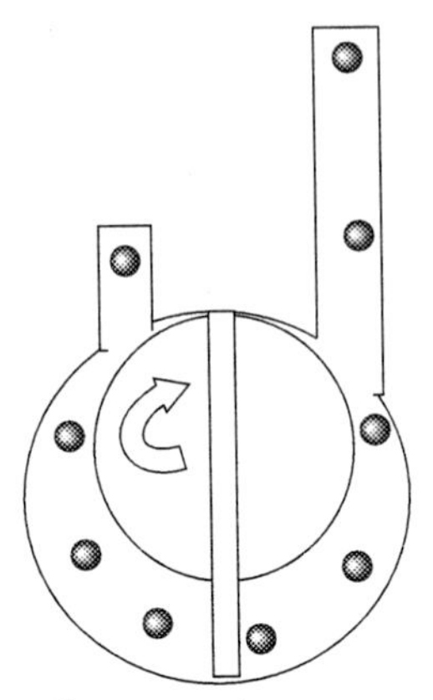

Compressed gas now exhausted while new gas molecules continue to enter the pump behind rotating vane

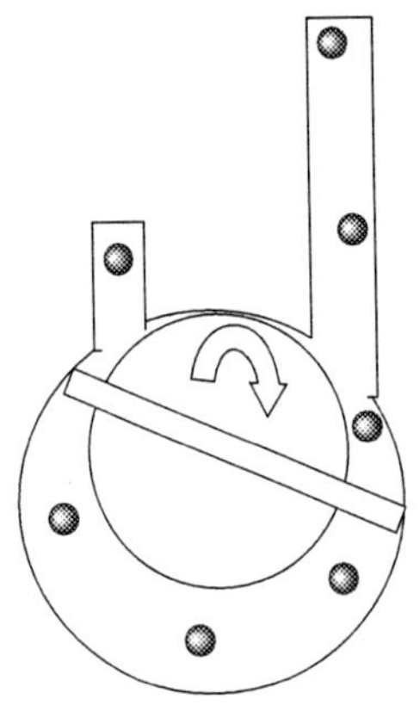

Compressed gas is completely exhausted while a new gas is trapped behind rotating vane, and the cycle repeats

FIGURE 4.7(b)

When oil-sealed mechanical pumps are operated at low pressures, they tend to *backstream* oil vapor back into the roughing line and the process chamber, and this may affect the process being run. Oil migration can be controlled by inserting traps, such as molecular sieve traps, in the roughing line. Another solution is to use oil-free mechanical pumps, such as diaphragm and scroll pumps.

4.4.2 Scroll Pump

Oil-free scroll pumps like the one shown in Figure 4.8 provide a clean operating environment with no sealant or lubricant required. They are ambient air–cooled and do not require cooling water.

FIGURE 4.8
Scroll pumps.
Source: (Varian Vacuum Technologies, *Product Catalog 2000,* p 16. Varian TriScroll Dry Vacuum Pumps)

Scroll pumps use two scrolls, or plates, shaped into a spiral called an *involute curve.* The scrolls fit inside one another. One scroll is fixed in place and remains stationary during operation. The other scroll moves with an orbital motion around the stationary scroll.

As shown in Figure 4.9, the orbiting motion of the movable scroll creates a crescent-shaped gas pocket, allowing gas from the inlet to move into the pump. This gas pocket is sealed, and then, as the movable scroll moves in its orbital trajectory, the trapped gas is compressed. The gas moves along the scroll toward the center of the scroll. When the gas reaches the center, it is expelled through the pump's outlet. The sequence of gas capture, gas compression, and gas exhaust is performed over and over again as the movable scroll moves in its orbital trajectory.

     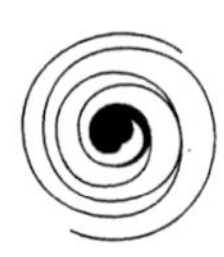

FIGURE 4.9
Pumping mechanism of a scroll pump.
Source: (Synergy Vacuum About Oil-Free Scroll Pumps, http://www.synergyvacuum.com/body02.htm)

Besides eliminating the risk of oil contamination and the need for oil traps and mist eliminators, the benefits of using scroll pumps include high pumping speed, compact size, and low ultimate pressure. Pumping speeds range from 300 to 600 liters per minute, ultimate pressures from 10^{-2} torr to 10^{-3} torr, and cost from $4,000 to $9,000. Scroll pumps can be configured for either single-phase or three-phase operation at 50 Hz or 60 Hz.

4.4.3 Diaphragm Pump

Diaphragm pumps are one of the oldest known pumps. They are simple in design and relatively low-cost.

When we breathe, we are actually using a diaphragm pump. As the diaphragm at the bottom of our chest cavity is pulled down, expanding the volume of the chest cavity, air is drawn into our lungs. When the diaphragm is pushed up, the volume of the chest cavity decreases, the air in our lungs is compressed, and we exhale. We repeat this pumping cycle over and over again throughout our life.

Figure 4.10 shows the pumping cycle for a diaphragm pump. A diaphragm made of a flexible material forms one wall of a small chamber. The flexible diaphragm is attached

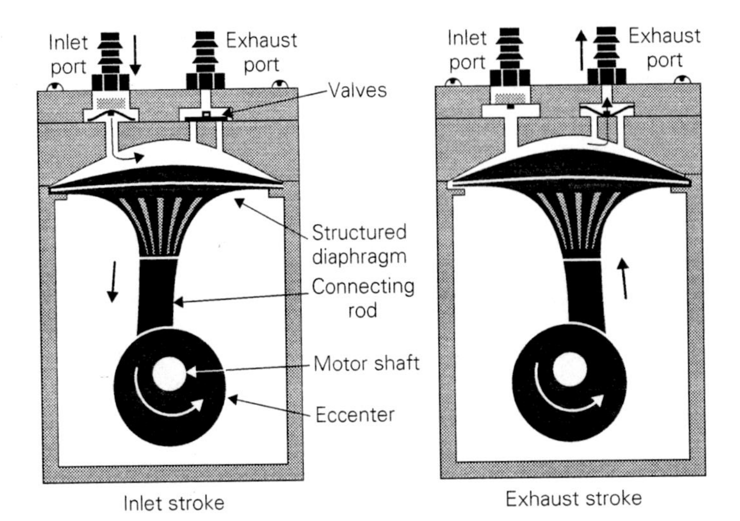

FIGURE 4.10
Pumping mechanism of diaphragm pump.
Reprinted with permission of KNF Neuberger USA, Trenton, NJ.

to a piston, which is attached to an eccentric cam driven by an electric motor. During the gas capture phase of operation, the piston rod is pulled down, causing the volume of the small chamber to increase. Gas is drawn in through the inlet port.

At a certain point in the rotation of the cam, the piston rod changes direction and is now pushed upward. This upward motion of the piston pushes the diaphragm up and reduces the volume of the small chamber. The trapped gas is now being compressed, and the pressure in the small chamber increases. The exhaust port opens and the gas is expelled from the chamber. This completes one pumping cycle.

Diaphragm pumps may be made with multiple chambers to improve pumping performance. Up to four chambers can be combined to provide ultimate pressures in the range of 0.5 torr. Pumping speeds range from 10 to 60 liters per minute. Most diaphragm pumps run at either 120 V or 220 V at either 50 Hz or 60 Hz. They range in price from $1,000 to $4,000.

4.4.4 Other Rough Vacuum Pumps

There are several other types of rough vacuum pumps. These include rotary piston pumps, claw pumps, screw pumps, and lobe pumps. Each of these pumps implements the gas capture, gas compression, and gas exhaust cycles using a different geometry. Detailed descriptions of these pumps will not be included here. Instead, a brief overview will be provided.

Rotary piston pumps are similar to rotary vane pumps. The eccentric cam drives a piston that is used to draw the gas into the pump, compress it, and exhaust it. Pumps of this kind are typically used as roughing pumps on large vacuum systems either alone or in combination with lobe blowers. They are rugged and mechanically simple, providing high pumping speeds.

The claw pump uses two claws rotating in opposite directions to capture, compress, and exhaust the gas. Claw pumps can be used to effectively pump corrosive and abrasive gases. Often they are combined with lobe pumps to achieve high pumping speeds and pressures in the millitorr range.

Another type of pump used to pump corrosive and abrasive gases is the screw pump. It finds application in backing turbomolecular pumps in reactive-ion etching and chemical vapor-deposition systems. The screw pump uses a pair of large rotating screws to move the gas from the inlet end of the pump to the exhaust end. Like the claw pump, screw pumps have high pumping speeds and ultimate pressures in the millitorr range.

Lobe pumps, also known as Roots pumps, use two figure-eight-shaped lobes or rotors to move the gas through the pump. The lobes spin in opposite directions creating two pumping sequences during each rotation at speeds of 3,000 to 3,500 rpm. In this pump, the gas is not compressed before it is exhausted.

4.5 ROUGH VACUUM GAUGES

There are a number of rough vacuum gauges that can measure pressure in the range of 760 torr to 1 millitorr. For our discussion in this text, two types of rough vacuum gauges will be described. First we will consider the various thermal conductivity gauges, which include thermocouple gauges, Pirani gauges, and convectorr gauges. Then we will discuss the capacitance diaphragm gauge, or capacitance manometer.

4.5.1 Thermal Conductivity Gauges

Thermal conductivity gauges are indirect-reading vacuum gauges. The term *indirect reading* implies that the gauge does not read pressure directly, but instead measures some other parameter associated with the gas and then converts it to a pressure reading. As their name implies, thermal conductivity gauges measure temperature and translate or convert the temperature reading to a pressure reading.

A gas has the ability to carry away heat. We have all gently blown on a hot morsel of food to cool it before putting it in our mouth. We are also familiar with Thermos bottles that use the absence of gas in the wall of the bottle to minimize heat transfer to or from the liquid inside the Thermos bottle. In this way, we can keep cold liquids cold and hot liquids hot.

To make a gauge that measures pressures in the rough vacuum regime, all we have to do is place a heated surface in the vacuum chamber and measure how much heat is lost from the surface. Heat is lost from the surface every time a gas molecule hits it and rebounds back into the chamber, taking some heat energy from the surface. The more collisions between gas molecules and the surface, the greater the heat loss, and therefore the higher the pressure. Conversely, the fewer collisions between gas molecules and the surface, the smaller the heat loss, and the lower the pressure. The amount of heat loss can be measured using either the Fahrenheit, Celsius, or Kelvin temperature scales.

Figure 4.11 shows the construction of a thermocouple gauge. The gauge controller applies a set voltage to the filament to heat it up to its operating temperature. A thermocouple is attached to the filament to sense the temperature of the filament. The gauge controller monitors the current produced by the thermocouple and produces a pressure reading that corresponds to the magnitude of the current.

4.5.2 Pirani Vacuum Gauge

A Pirani gauge, shown in Figure 4.12, is a thermal conductivity gauge that uses the change in resistance of the filament in the gauge as one arm of a Wheatstone bridge. A *Wheatstone bridge* in its simplest form consists of four resistances arranged in a square or diamond shape, like a baseball diamond. Using the baseball analogy, a meter is placed between first and third base and a power supply between home plate and second base.

The circuit, shown in Figure 4.13, operates in the following manner. If the bridge is balanced, the voltage at first base equals the voltage at third base, and there is no voltage potential across the meter. No voltage means no current flows, and the meter reads zero. If one of the resistances changes, for example the resistance of the Pirani transducer, due to a change of pressure in the vacuum chamber, resulting in a change in heat loss from the filament, the bridge will become unbalanced. The voltages at first and third base will not be equal, and the potential difference across the meter will cause a current to flow through the meter. If the meter has an analog readout, the current flowing through the meter will cause the meter needle to deflect and show a non-zero reading. The magnitude of the deflection will be proportional to the new pressure in the vacuum chamber.

Using a thermal conductivity gauge is not as simple as hooking up the gauge to the chamber or lines in a vacuum system and reading the numbers off the gauge controller. Since the thermal conductivity of different gases varies, the number representing the pressure may

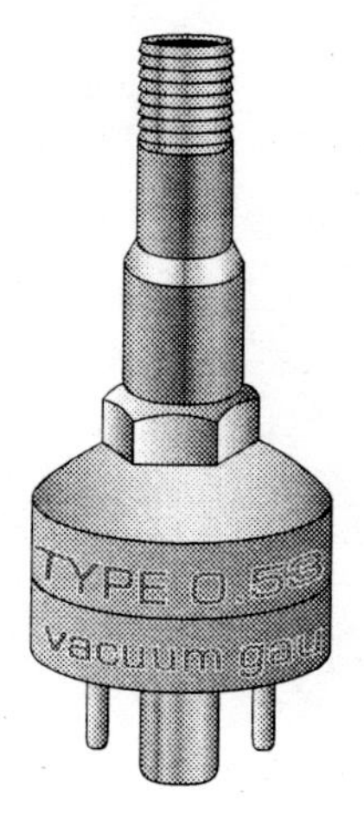

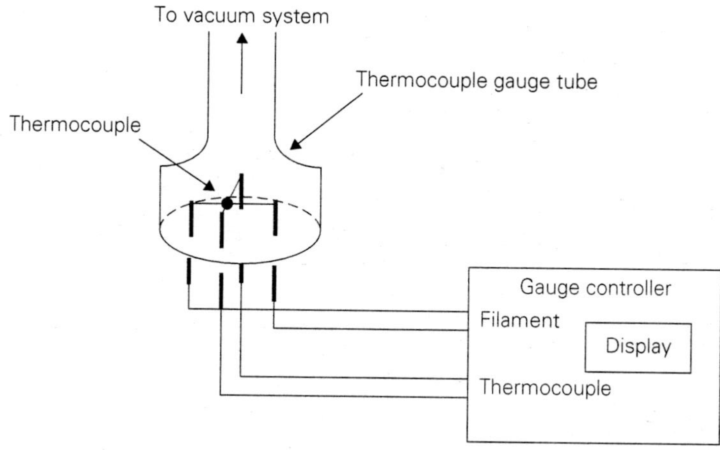

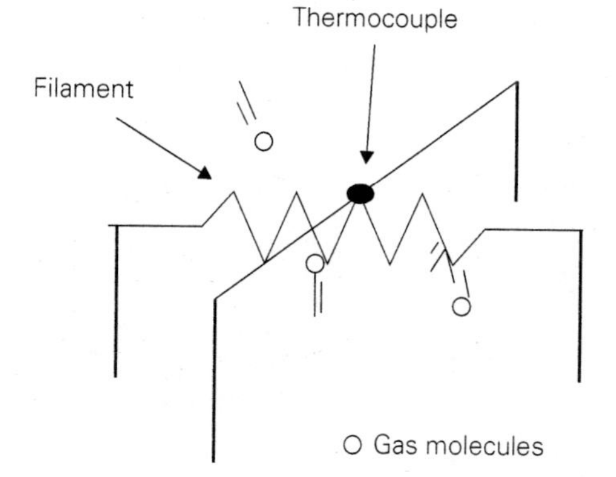

FIGURE 4.11
Thermocouple gauge and diagram of its internal structure.
Source: (Varian Vacuum Technologies, *Product Catalog 2000,* p. 307. 531 Thermocouple Gauge)

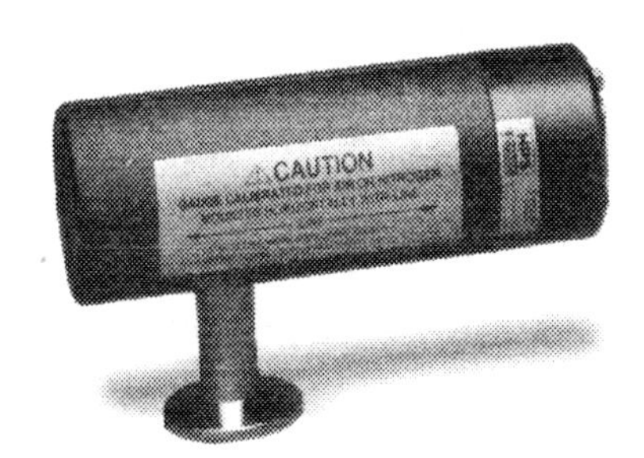

FIGURE 4.12
Pirani gauge.
MKS Series 947 Digital Convection-Enhanced Pirani Vacuum Sensor System (Data Sheet)

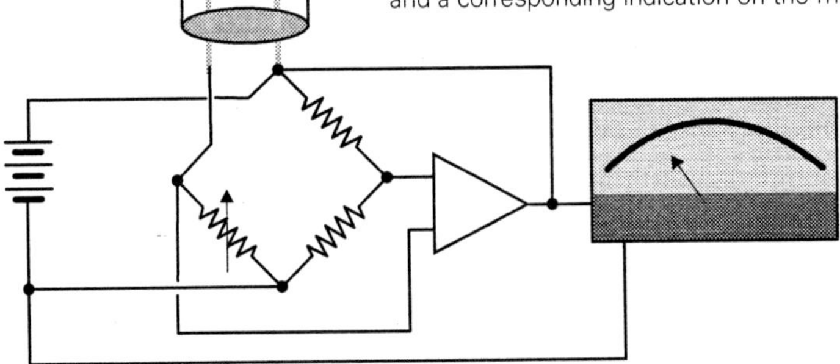

FIGURE 4.13
Circuit diagram of bridge circuit for a Pirani gauge.
Source: MKS Instruments Introduction to Vacuum Gauging Techniques, p. 6.

TABLE 4.1
Thermal conductivity of common gases

Gas	Thermal conductivity MJ/(s-m-K)
Air	24.0
Argon	16.6
Carbon Dioxide	14.58
Helium	142.0
Hydrogen	173.0
Methane	30.6
Neon	45.5
Nitrogen	24.0
Oxygen	24.5
Water Vapor	24.1
Xenon	4.50

have to be adjusted if the gas composition is different from the gas composition for which the gauge is calibrated. Table 4.1 gives the thermal conductivity of common gases.

What Table 4.1 tells us is that the amount of heat lost from the filament in a thermal conductivity gauge depends on the gas in the system. For example, if the chamber is filled with helium as opposed to nitrogen, the helium, having a higher thermal conductivity, will carry more heat away from the filament than nitrogen, and thus an equivalent amount of helium will cool the filament to a greater extent than nitrogen. Argon, on the other hand, with a lower thermal conductivity, will not cool the filament as fast as nitrogen. So with any gas other than nitrogen, you should not take the observed pressure reading at face value.

Let's use the PG 105 convention-enhanced Pirani gauge produced by Stanford Research Systems as an example. All PG 105 gauges are factory-calibrated and temperature-compensated for nitrogen (air). The response of the gauge is very well characterized, and with proper calibration data it is possible to obtain accurate pressure measurements for other gases as well. For user convenience, nitrogen- and argon-specific calibration curves are loaded into SRS's IGC 100 gauge controller, making direct pressure measurements possible for these gases. For other gases, gas-correction curves are provided for the PG 105 gauge (see Figure 4.14). However, for gases, or mixtures of gases, not included in the gauge data sheet, users will have to generate their own conversion curves.

For the PG 105 convection-enhanced Pirani gauge, the actual pressure is equal to the nitrogen-equivalent reading from the gauge controller times the gas-correction factor. For pressures below 1 torr, the gas correction factor for argon is given as 1.59. Hence, if the pressure reading on the controller display is 0.50 torr (the nitrogen-equivalent reading), then the actual pressure is equal to 0.50 torr $\times$ 1.59, or approximately 0.8 torr. Controllers such as the IGC 100 will perform this calculation for you and display the actual argon pressure on the controller readout.

On the other hand, if the vacuum chamber is filled with methane, and the observed pressure reading on the gauge controller is 0.50 torr, we can use the given gas-correction figure for methane to determine the actual pressure. From the data sheet, the gas-correction figure for methane is 0.63. Multiplying the gas-correction figure, 0.63, by the gauge pressure reading, 0.50 torr, we obtain 0.32 torr, the actual pressure for methane in the chamber.

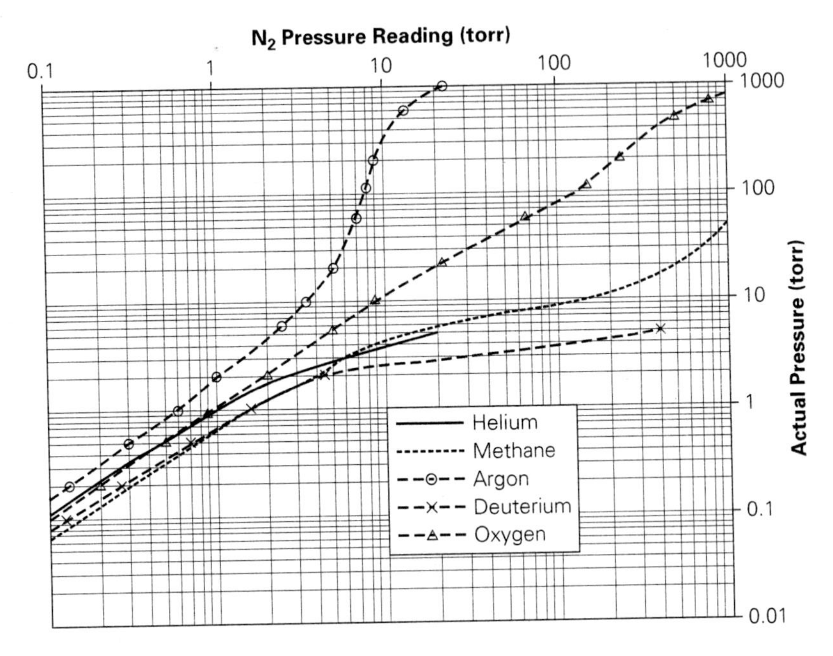

FIGURE 4.14
PG-105 gauge indicated pressure (N2 equivalent) vs. actual pressure curve.

It is important to read the operating instructions and data sheet for the pressure gauge you are using. In some cases, you may be instructed to divide (and not multiply) the observed pressure reading by the gas-correction factor.

4.5.3 Capacitance Diaphragm Gauge

A capacitance diaphragm gauge, or capacitance manometer, is a direct reading gauge. It senses the deflection of a flexible metal diaphragm in the gauge produced by the collective force of many gas molecules striking the diaphragm. The metal diaphragm forms one plate of a parallel plate capacitor. Since $C = \epsilon \frac{A}{d}$ where C is the capacitance in farads, ε is the permittivity of the dielectric material, A is the plate area, and d is the plate separation, capacitance is inversely proportional to the separation of the two plates forming the capacitor, because the plate area and the dielectric material do not change. The electronic circuitry in the gauge controller converts this change in capacitance to a corresponding frequency and then to a pressure readout. Figure 4.15 shows a capacitance manometer.

Capacitance manometers are used as process-control monitors because they are gas species independent and very accurate. They operate over four decades of pressure from 1,000 torr to 0.1 torr. They are very fast, with response times on the order of 1 millisecond

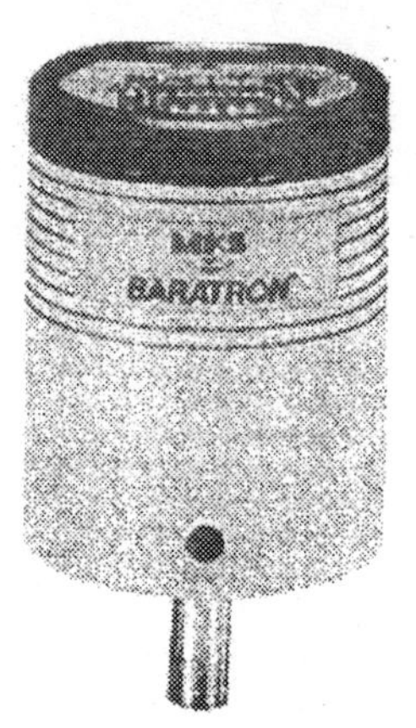

FIGURE 4.15
Capacitance manometer.

or less, and have an accuracy of ± 0.25 % of full scale. The down side of using capacitance manometers is that they are more expensive than other gauges within the same pressure range, such as the thermocouple and Pirani gauges.

4.6 VALVES AND FITTINGS

Given a chamber and a pump, other vacuum system components are needed to configure a suitable vacuum system. Piping is needed to connect the pump to the chamber, and usually a valve is inserted in this line to provide isolation. A pressure gauge can be used to monitor chamber pressure, and a back-to-air valve is needed to vent the chamber back to atmospheric pressure.

At pressures above 10 millitorr, gas molecules act much like a fluid. We can use our familiarity with the flow of liquids to help us visualize what is happening in a vacuum system in the rough vacuum regime. Flow in this pressure regime is known as *viscous flow.* The term "viscous" implies that the gas molecules are "thick," so thick that they are constantly bumping into each other and the walls of the chamber. In fact, the molecules are so closely packed that when some of them are pumped out of the chamber, other gas molecules will rush to fill up the empty space and distribute themselves within the chamber.

Molecular movement in the viscous flow regime is very predictable. Because of this predictability, smaller-diameter hoses and pipes can be used while moving large quantities of molecules per unit time from one place to another. Similarly, a garden hose with a diameter of a half-inch works fine at the water pressures of the city water system.

In the rough vacuum regime, a wide variety of materials can be used for piping between system components. For example, PVC cord–reinforced flexible vacuum hose is an easy way to connect a vacuum pump to a chamber. In a manufacturing setting, however, the piping is usually stainless steel or aluminum.

Klamp Flanges (KF) are common class of vacuum flanges and fittings used in the rough vacuum regime. As shown in Figure 4.16, KF flanges use a centering ring with an elastomer O-ring as the sealant. The O-rings are made of either Buna-N or Viton. The O-ring is placed between the two components to be joined. A clamp is then used to hold the

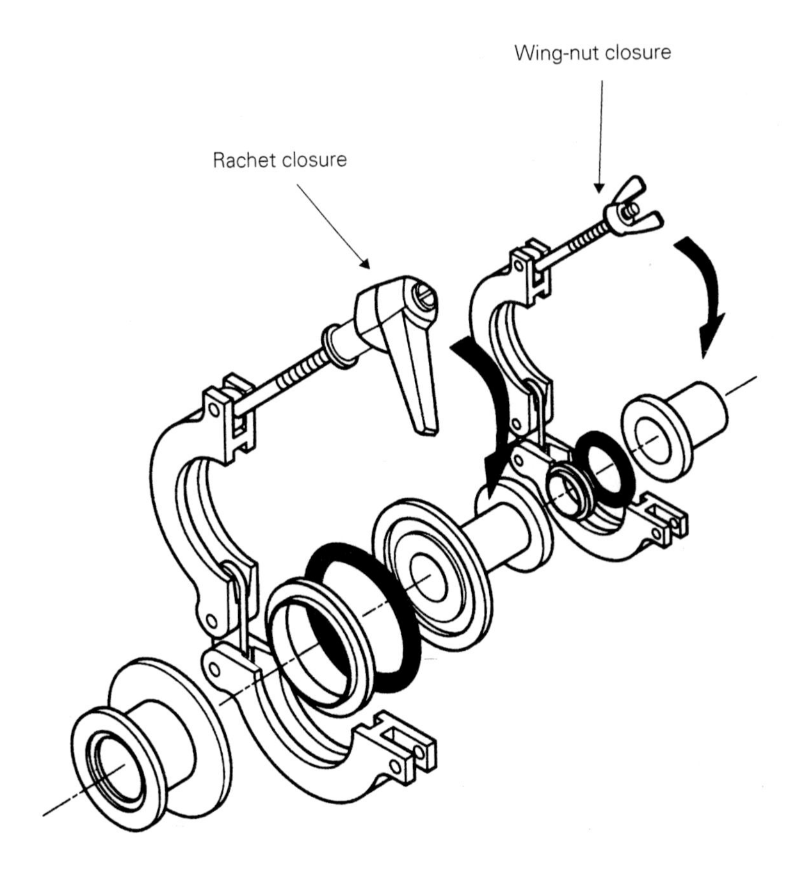

FIGURE 4.16
KF flanges.
Source: Varian Vacuum Products, *Basic Vacuum Practice*, 3rd Edition, p. 159

two components together and compress the elastomer O-ring to form a good seal. Flanges come in standard sizes ranging from NW10, the smallest size, to NW50, the largest size.

When using KF flanges to make a connection, it is important to ensure that the groove and mating surface are clean and dry. Check the sealing surfaces for scratches that cross the sealing area. If O-rings are reused, visually inspect them to make sure they do not have small cross-wise cracks or nicks that might leak. If an O-ring has been exposed to solvents or excessive heat that has caused it to swell, do not reuse it. Replace it with a new one.

Valves are used to provide isolation between vacuum components. The block valve is commonly used for this purpose in the rough vacuum regime. Block valves are made of aluminum, stainless steel, or brass. They can either be hand-operated, air-operated, or activated electromagnetically. Flange options include KF flanges as well as others. They can be either in-line or right-angle. Figure 4.17 shows some examples of aluminum block valves with KF flanges.

Klamp Flange fittings include nipples, elbows, tees, and reducers, to name a few. Figure 4.18 shows some of these KF fittings.

FIGURE 4.17
Aluminum block valves with KF flanges.
Source: Varian Vacuum Technologies, *Product Catalog 2000,* p. 334.

FIGURE 4.18
KF fittings.
Source: Varian Vacuum Technologies, *Product Catalog 2000.*

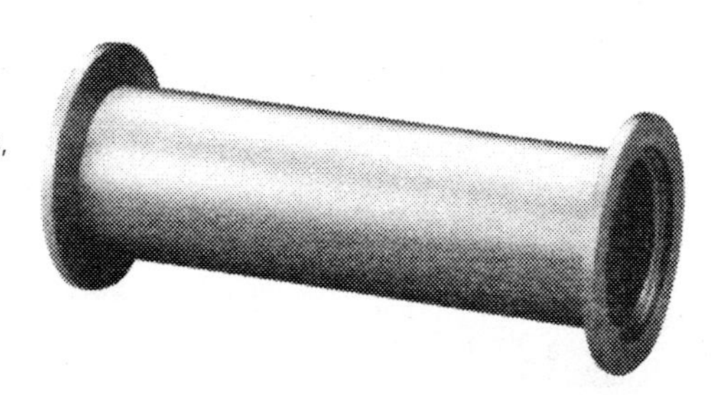

Nipples

Elbows

Tees

And finally, we need a valve that we can use to vent the system back to atmospheric pressure. Stainless steel, elastomer-sealed valves can be used for this purpose. They can also be used as gas inlets for process gases to the process chamber.

4.7 ROUGH VACUUM PUMP DOWNS

For pressures above 10^{-2} torr, the volume of the chamber and pumping speed are the determining factors when calculating pump-down times. Let's assume that we have a chamber of volume V and a pump connected directly to the chamber having an effective pumping speed S_{eff}. The ultimate pressure that can be obtained for the given pump and chamber shall be P_{ult}. At time $t = 0$, the initial pressure shall be $p = P_i$. The pump-down time from P_i to some final pressure P_f is given by

$$t = \frac{V}{S_{eff}} \ln \frac{P_i - P_{ult}}{P_f - P_{ult}}. \tag{4.1}$$

Equation 4.1 assumes that the chamber is clean and does not have any leaks. It also assumes that the conductance of the connection between the pump and the chamber is much greater than the effective pumping speed of the pump.

Equation 4.2 implies that the pressure versus time curve follows an exponential decay of the form

$$p(t) = (P_i - P_{ult})e^{\left(-\frac{S_{eff}}{V}t\right)} + P_{ult}. \tag{4.2}$$

EXAMPLE 4.1 A Varian SD-201 rotary vane pump with a nominal pumping speed of 193 liters per minute is used to evacuate a 300-liter chamber. How long will it take to pump down the chamber from atmosphere to 1 torr?

Solution

From the problem statement, we are given the volume of the chamber, V, as 300 liters, the effective pumping speed, S_{eff}, as 193 liters per minute, the starting pressure, P_i, as atmospheric pressure, which is assumed to be 760 torr, and the final pressure, P_f, as 1 torr. From the data sheet for the SD-201, the ultimate vacuum pressure, P_{ult}, is 3 millitorr. Substituting this information into Equation 4.1 yields

$$t = \frac{300\ \text{L}}{193\ \text{L/min}} \ln \frac{760\ \text{torr} - 0.003\ \text{torr}}{1\ \text{torr} - 0.003\ \text{torr}}$$

Since 0.003 torr is much smaller than 1 torr, the equation simplifies to

$$t = \frac{300\ \text{L}}{193\ \text{L/min}} \ln \frac{760\ \text{torr}}{1\ \text{torr}}$$

$$t = 1.55\ \text{min} \ln(760)$$

$$t = 10.31\ \text{min}.$$

EXAMPLE 4.2 A Varian MD-40 Mechanical Diaphragm Pump is used to evacuate a 20-liter chamber. How long will it take the MD-40 to reduce the pressure in the chamber from atmosphere to 10 torr?

Solution

From the problem statement, the volume of the chamber, V, is 20 liters, the initial pressure, P_i, is atmospheric pressure, or 760 torr, and the final pressure, P_f, is 10 torr. From the MD-40 data sheet, the nominal pumping speed for the MD-40 is 40 liters per minute, and the ultimate vacuum is less than 3.8 torr.

Substituting this information into Equation 4.1 yields

$$t = \frac{15 \text{ L}}{40 \text{ L/min}} \ln \frac{760 \text{ torr} - 3.8 \text{ torr}}{10 \text{ torr} - 3.8 \text{ torr}}$$

Performing the computations yields

$$t = 0.38 \text{ min} \ln(122)$$
$$t = 1.83 \text{ min.}$$

Note that in this case the ultimate vacuum pressure may be significant. If the ultimate pressure value is assumed to be negligible, the pump-down time calculation yields 1.65 minutes, about a 10% difference.

EXAMPLE 4.3 Calculate the pump-down time for a Varian SD-2500 Mechanical Pump to reduce the pressure in a 200-cubic-foot chamber from atmosphere to 0.1 torr. Use the pump speed curve for the SD-2500 shown in Figure 4.19.

Solution

From the SD-2500 pumping speed curve, we observe that the pumping speed is approximately 70 cfm from atmosphere to about 3 torr. Below 3 torr, the pumping speed rolls off with a slope of 25–30 cfm per decade of pressure. If we assume that the pumping speed is a constant 70 cfm over the entire pressure range from atmosphere to 0.1 torr, our calculated pump-down time will be shorter than the actual pump-down time. We need a better method for calculating the pump-down time.

One alternative method is to use a piece-wise linear approximation for the SD-2500 pumping speed curve. To implement this method, let us divide the pressure range into two segments, namely atmosphere to 1 torr, where the pumping speed is a constant 70 cfm, and 1 torr to 0.1 torr, where the pumping speed decreases from 70 cfm to approximately 47 cfm at 0.1 torr. We can find an average pumping speed in the 1 torr to 0.1 torr range by adding 70 cfm and 47 cfm and dividing by 2. This yields an average pumping speed over this range of 58.5 cfm. Now we can calculate the pump-down time for each pressure segment and then add the two time segments together to find the total pump-down time.

$$t_{pd} = \frac{200 \text{ cfm}}{70 \text{ cfm}} \ln\left(\frac{760 \text{ torr}}{1 \text{ torr}}\right) + \frac{200 \text{ cfm}}{58.5 \text{ cfm}} \ln\left(\frac{1 \text{ torr}}{0.1 \text{ torr}}\right)$$

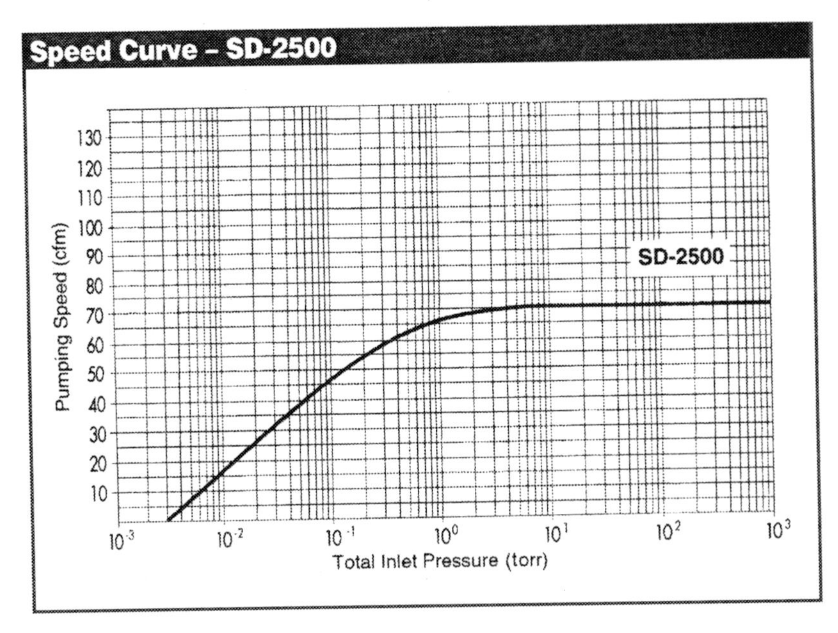

FIGURE 4.19
Varian SD-2500 mechanical pump speed curve.

$$t_{pd} = 18.91 \text{ min} + 7.87 \text{ min}$$

$$t_{pd} = 26.78 \text{ min}$$

How does this compare to the pump-down time using the assumption that the pumping speed is a constant 70 cfm? The pump-down calculation using the constant 70 cfm pumping speed yields a pump-down time of 25.53 minutes, a difference of 1.25 minutes. Is this significant? It depends on the application. But from a purely time perspective, a 5% difference doesn't seem like that much.

4.8 CONDUCTANCE

But wait a minute. What about the effect of the piping if the rough pump has to be placed at some distance from the vacuum chamber? Wouldn't it affect the pump's ability to remove gas from the chamber? The answer is, It depends.

From our experience, we know that pipes or tubes of varying lengths and cross-sections can conduct gases and liquids with different degrees of difficulty. Say you are drinking a soft drink through a straw. Drawing the liquid through a regular 8-inch straw poses no great difficulty. Now imagine that the straw is 30 inches long. Although I have not actually tried it, I imagine that it would take more effort to drink the liquid if one had to draw it along that length of straw. Or imagine that an 8-inch straw of one-tenth the diameter is used. Again, I have not actually tried this, but I imagine that it would again require more effort to draw the liquid through the narrower straw.

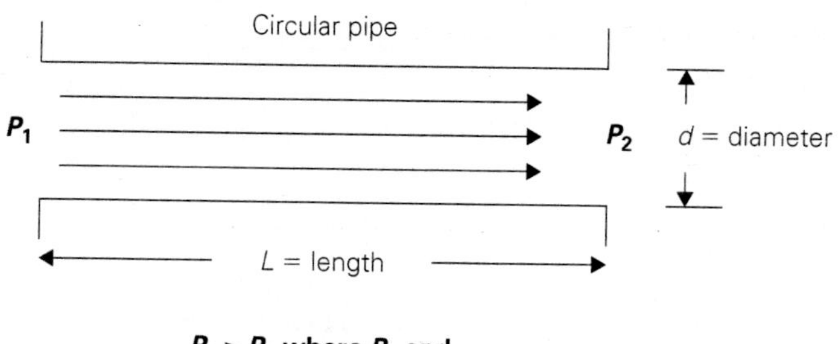

FIGURE 4.20
Cross-sectional view of pipe carrying gas.

The ease with which a gas or liquid is drawn through a pathway of given geometry is quantified by a property called *conductance*. Conductance values depend not only on the pressure and the nature of the gas, but also on the shape of the conducting element. Other factors are the conducting element's length and whether it is straight or curved. The result is a large set of equations that take into account various combinations of these elements that affect conductance.

For example, the conductance of a straight pipe or tube (C_t) shown in Figure 4.20, which is not too short, of length L and with a circular cross-section of diameter d, operating in the laminar flow range, is given by

$$C_t = 180 \frac{\mathrm{L}}{\mathrm{cm}^3\ \mathrm{torr\ sec}} \frac{(d)^4}{L} (p_{ave}) \tag{4.3}$$

where C_t is the conductance in liters per second,
d is the cross-sectional diameter of the pipe in centimeters,
L is the length of the pipe in centimeters,
and, p_{ave} is the average pressure in the pipe in torr,

$$p_{ave} = \frac{p_1 + p_2}{2}$$

where p_1 is the pressure at the start of the pipe (along the direction of flow), and p_2 is the pressure at the end of the pipe.

EXAMPLE 4.4 What is the conductance of a 15-inch pipe with an inside diameter (ID) of 0.5 inches if the pressure is 20 torr at the start of the pipe and 1 torr at the end of the pipe?

Solution

We are given the following information:

$$L = 15 \text{ inches} = 15 \text{ inches} \times 2.54 \text{ cm/in} = 38.1 \text{ cm}$$
$$d = 0.5 \text{ inches} = 0.5 \text{ inches} \times 2.54 \text{ cm/in} = 1.27 \text{ cm}$$

P_{ave} can be calculated from the pressures at the ends of the pipe,

$$p_{ave} = 10.5 \text{ torr.}$$

Substituting these values into Equation 4.3 yields,

$$C_t = 180\frac{\text{L}}{\text{cm}^3 \cdot \text{torr} \cdot \text{sec}} \cdot \frac{(1.27\ \text{cm})^4}{38.1\ \text{cm}} \cdot (10.5\ \text{torr})$$

$$C_t = 129\frac{\text{L}}{\text{sec}}$$

Equation 4.3 can only be used under viscous and laminar flow conditions. How can we tell if we are in viscous flow? The following condition, based on the ultimate pressure, P_u, and the cross-sectional diameter, d, of the circular pipe, can be used,

$$P_u \cdot d > 5 \times 10^{-1}\ \text{torr} \cdot \text{cm}.$$

For the preceding example, the ultimate pressure was 1 torr and the diameter was 1.27 cm. Applying the test for viscous flow,

$$P_u \cdot d = 1\ \text{torr} \cdot 1.27\ \text{cm} = 1.27\ \text{torr} \cdot \text{cm} > 5 \times 10^{-1}\ \text{torr} \cdot \text{cm}.$$

Thus the condition for viscous flow is met, and the use of Equation 4.3 is valid.

We now ask a follow-up question: Is there a conductance at the aperture where the gas enters the pipe from the chamber? (See Figure 4.21.) There certainly seems to be some congestion occurring here, much like the masses of people leaving a football stadium at the end of the game.

For viscous flow, the conductance of the aperture can be calculated using the following equation,

$$C_a = 20\frac{\text{L}}{\text{cm}^2 \cdot \text{sec}} \frac{A}{1 - \delta} \tag{4.4}$$

where C_a is the conductance of the aperture in liters-per second,
A is the cross-sectional area of the aperture in cm^2,

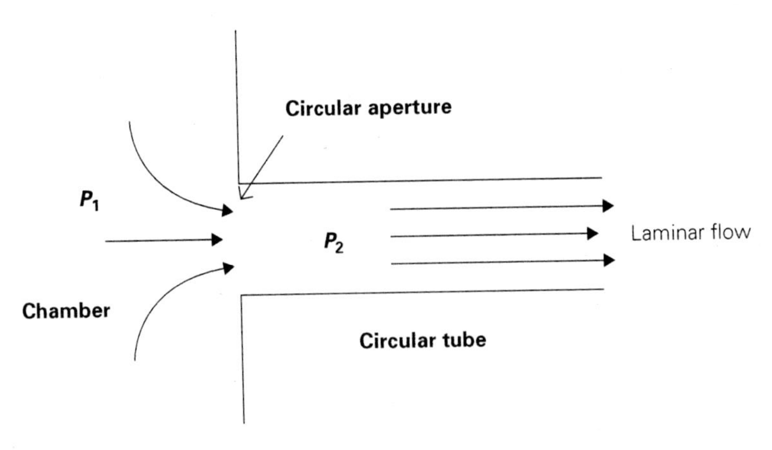

FIGURE 4.21
Cross-sectional view of chamber, aperture, and tube.

and, δ is the ratio of P_2 to P_1 where P_2 is the pressure in the pipe, and P_1 is the pressure in the chamber. For viscous flow, δ is very small, and Equation 4.4 reduces to

$$C_a = 20\frac{\text{L}}{\text{cm}^2 \cdot \text{sec}}A \tag{4.5}$$

Note that in the viscous flow range the conductance is not a function of pressure, but only depends on the geometry of the aperture.

Let's return to our previous example and determine the conductance of an aperture of diameter 0.5 inch, or 1.27 cm. Using Equation 4.5 yields

$$C_a = 20\frac{\text{L}}{\text{cm}^2 \cdot \text{sec}}(\pi \cdot (1.27\text{ cm})^2)$$

$$C_a = 101.3\frac{\text{L}}{\text{sec}}$$

The conductance of the tube and the conductance of the aperture act as series elements. Thus the two conductance can be combined as follows:

$$\frac{1}{C_{net}} = \frac{1}{C_t} + \frac{1}{C_a} \tag{4.6}$$

or

$$C_{net} = \frac{C_t \cdot C_a}{C_t + C_a}. \tag{4.7}$$

Substituting our values for our 15-in. pipe of 0.5-in. diameter yields a net conductance of

$$C_{net} = \frac{129\frac{\text{L}}{\text{sec}} \cdot 101.3\frac{\text{L}}{\text{sec}}}{129\frac{\text{L}}{\text{sec}} + 101.3\frac{\text{L}}{\text{sec}}}$$

$$C_{net} = 56.7\frac{\text{L}}{\text{sec}}$$

The final step is to combine the conductance of the piping and the conductance of the vacuum pump. Since the conductance of the piping and the conductance of the pump are in series, the net pumping speed can be found using

$$\frac{1}{S_{net}} = \frac{1}{C_{net}} + \frac{1}{S_p} \tag{4.8}$$

where S_{net} is the net conductance or pumping speed of the pump plus the piping,
C_{net} is the net conductance of the piping,
and, S_p is the pumping speed of the vacuum pump.

Equation 4.8 can be rearranged, and an equation for S_{net} can be written,

$$S_{net} = \frac{C_{net} \cdot S_p}{C_{net} + S_p} \tag{4.9}$$

EXAMPLE 4.5 What is the net pumping speed of a pumping system consisting of a 90-liter-per-minute mechanical pump connected to a vacuum chamber through piping having a conductance of 60 liters per second? Give the net pumping speed in liters per second.

Solution

Since the net pumping speed is to be specified in liters per second, the pumping speed of the mechanical pump should be converted to liters per second.

$$S_p = 90\frac{\text{L}}{\text{min}} \times \frac{1\ \text{min}}{60\ \text{sec}} = 1.5\frac{\text{L}}{\text{sec}}$$

Substituting the given values for C_{net} and S_p into Equation 4.9 yields

$$S_{net} = \frac{\left(60\frac{\text{L}}{\text{sec}}\right) \cdot \left(1.5\frac{\text{L}}{\text{sec}}\right)}{\left(60\frac{\text{L}}{\text{sec}}\right) + \left(1.5\frac{\text{L}}{\text{sec}}\right)}$$

$$S_{net} = 1.46\frac{\text{L}}{\text{sec}}$$

In this example, the conductance of the piping is much greater than the pumping speed (conductance) of the pump. Therefore the factor limiting the net pumping speed is the pumping speed of the mechanical pump, not the size of the piping. However, this is not always the case.

4.9 TROUBLESHOOTING ROUGH VACUUM SYSTEMS

There are a number of things that can go wrong with a rough vacuum system. Knowing your vacuum system helps in detecting and troubleshooting system problems. This section discusses some common problems and how to fix them.

No Start-Up

Failure to start up is one of the simpler things to diagnose and possibly to fix. The observed symptom is the failure of the mechanical pump to start when the power is turned on. The cause may be that power is not getting to the pump. Corrective action begins with checking the electrical system. This includes checking the power switch to see if it is actually in the ON position, and checking the circuit breaker, which may have switched OFF. Another possibility is an opening in the power line running to the pump. A voltmeter can be used to measure the voltage at the pump.

Another possibility is that the pump has frozen or seized up. In this case, the observed symptom may be a whining or buzzing noise indicating that the motor is trying to turn, but cannot do so for some reason. If this condition persists, overheating may be detected by a burning odor.

Pump is running but no vacuum is detected

If the pump is operating normally but no vacuum is detected in the chamber, there are several possible causes. If the mechanical pump uses a belt drive, it is possible that the belt has broken or slipped off the pulley(s). Another possibility is that the foreline valve has not been opened. A third possibility is that the vent valve has not been closed and is in the open position.

It is also possible that an accurate pressure reading is not being displayed. If a multi-gauge controller is being used, it may be displaying a reading from the wrong gauge, a gauge in another part of the vacuum system that is looking at atmospheric pressure.

Deviation from standard pump-down curve

A deviation from the standard pump-down curve can be caused by an increase in the gas load or a change in the characteristics of the pressure gauge. Increases in the gas load can be caused by contamination in the system or in the vacuum pump, such as condensed water vapor, solvents, or gases in the pump oil.

Pressure readings that are too high or too low can be caused by a dirty or contaminated sensor. Often, the simplest solution in such cases is to replace the sensor.

Cannot reach base pressure

Failure to reach base pressure can be caused by an increase in gas load or a problem within the pump. Increased gas load often results from a leak, but this is not the only cause. Contamination can also increase the gas load significantly so that an increase in the base pressure is observed.

SUMMARY

Rough vacuum systems are relatively simple and minimally consist of a single mechanical pump, chamber, and pressure gauge, one or more valves, and piping. The types of mechanical pumps in use include rotary vane and piston pumps, diaphragm pumps, and scroll pumps. These pumps operate by expanding the volume, trapping a volume of gas, compressing the gas, and finally exhausting the gas from the pump. Pump selection will depend on the application, cost, pumping speed requirements, and space, among other factors.

Vacuum gauges used in rough vacuum systems can be either thermal conductivity pressure gauges, such as thermocouple and Pirani gauges, or direct-reading pressure gauges like capacitance manometers. Pressure gauge selection will be influenced by such factors as cost, pressure range, and response time. It is important to remember that thermal conductivity gauges are sensitive to gas type. They are calibrated for nitrogen, but gauge readings for other gases require correction using the correction factors supplied with the pressure gauge. Direct-reading pressure gauge readings do not require correction.

A wide variety of fittings can be used in rough vacuum systems. Common fittings include Klamp Flange (KF) fittings and ISO fittings. Conflat fittings can also be used in rough vacuum systems, but generally are not. Piping can be metal pipe or PVC cord-reinforced flexible vacuum hose with steel worm-gear hose clamps. Again it depends on the application, cost constraints, and other factors relevant to your application.

Rough vacuum pump downs will help characterize your rough vacuum system. By comparing present pump-down curves to past pump-down curves, you can detect changes in system performance and take corrective action, if necessary.

Maintaining rough vacuum systems is relatively easy. Mechanical vacuum pumps require regular maintenance to replace seals and diaphragms. Chambers may require periodic cleaning, depending on what is being pumped.

Troubleshooting rough vacuum systems usually focuses on leaks and worn-out system components. Failure of the vacuum system to pump down to base pressure can be caused by leaks, failure to close a manual vent valve, or the failure of an automatic vent valve to close. It can also be caused by wear on system components and may indicate the need for system maintenance to replace worn-out components.

BIBLIOGRAPHY

Basic Vacuum Practice, 3rd Edition, Varian Vacuum Products, Lexington, MA, 1992.

"Gas Correction Curves for PG105 Gauges," Stanford Research Systems, Sunnyvale, CA.

Hoffman, Dorothy M., Singh, Bawa, and Thomas, John H., III, Editors. *Handbook of Vacuum Science and Technology*, Academic Press, 1998.

"HPS Products Series 325 Moducell: Pirani Vacuum Sensor/Transducer Operation and Maintenance Manual," MKS Instruments, Wilmington, MA, 1999.

"HPS Products Series 907: Analog Convection Transducer (ACT) Vacuum Sensor Operation and Maintenance Manual," MKS Instruments, Wilmington, MA, 1999.

http://www.synergyvacuum.com/body02.htm

"Introduction to Vacuum Gauging Techniques," MKS Instruments, Wilmington, MA,1998.

Lafferty, J. M., Editor. *Foundations of Vacuum Science and Technology*, Wiley, Hoboken, NJ, 1998.

O'Hanlon, John. *A User's Guide to Vacuum Technology*, 3rd Edition, Wiley Interscience, Hoboken, NJ, 2003.

Product Catalog 2000. Varian Vacuum Technologies, Lexington, MA, 2000.

PROBLEMS

1. A 0.50-liter piston pump is connected to a 5-liter chamber. Assume that the pressure in the chamber is atmospheric pressure (1 atm). What will be the pressure in the chamber after eight strokes of the piston pump?

2. A 0.25-liter piston pump is connected to a 20-liter chamber. How many strokes of the piston pump will be needed to reduce the pressure in the chamber from one atmosphere to 0.75 atmospheres?
3. A MDP-30 diaphragm pump is connected to a cylindrical chamber 15 inches in diameter and 12 inches in length. How long will it take the MDP-30 to reduce the pressure in the chamber from atmosphere to 10 torr? The nominal pumping speed of the MDP-30 is 30 liters per minute. Assume that the pipe connecting the pump to the chamber is not a limiting factor.
4. A Varian SD-450 mechanical pump is connected to a 200-liter chamber. Assuming that the conductance of the connecting pipe is not a limiting factor, how long will it take to pump down the chamber from atmosphere to 0.5 torr? Show all calculations, and list all assumptions made.
5. Three identical thermal conductivity gauges are used to measure the gas pressure in three identical tanks of equal volume. Each tank contains the same number of moles of gas. Tank 1 contains dry nitrogen gas, Tank 2 contains argon gas, and Tank 3 contains helium gas. Which gauge will read the highest pressure, and why? Which tank will read the lowest pressure, and why?
6. Determine the net conductance of a circular pipe 25 cm long and 3.0 cm in diameter if the pressure at one end of the pipe is 30 torr and at the other end is 4 torr.

LABORATORY ACTIVITY

Rough Vacuum Pump Down

Equipment Needed: a rough vacuum pump, rough vacuum pressure gauge, block valve, and vent valve configured as a rough vacuum system; stopwatch.

Objective: To obtain a pump-down curve for a rough vacuum system.

Pre-Lab Activity: Determine the approximate volume of the vacuum chamber and attached volumes open to the chamber. Look up the pumping speed of the rough vacuum pump. Calculate the pump-down time from atmosphere to 1 torr.

Laboratory Procedure: The chamber has been vented to the atmosphere and is at atmospheric pressure.

1. Close the vent valve and roughing valve (block valve between the chamber and the vacuum pump).
2. Calibrate rough vacuum gauges for atmospheric pressure (760 torr).
3. Start the rough vacuum pump. This will pump down the volume between the vacuum pump and the block valve. The chamber pressure will not change.
4. Determine the time intervals at which chamber pressure readings will be made and construct a data sheet on which pressure readings will be recorded. Reset the stopwatch to zero.
5. Open the roughing valve.
6. Record pressure readings at the predetermined time intervals and enter them on your data sheet. Continue until base pressure is reached.

7. To shut down the vacuum system, close the roughing valve and turn off the rough vacuum pump. The chamber may be vented to the atmosphere or left under vacuum.

Analysis

Plot the pump-down curve (pressure versus time graph) for your vacuum system. Compare the pump-down time from atmosphere to 1 torr from your graph to the calculated pump-down time you obtained in the Pre-Lab Activity. If the measured and calculated pump-down times are significantly different, give a reason(s) to account for this difference.

CHAPTER

5

The High-Vacuum Regime

5.1 INTRODUCTION

In Chapter 3, we defined the rough vacuum regime as pressures ranging from 759 torr to 1 millitorr, about six orders of magnitude for pressure below atmosphere. The next pressure range is the high-vacuum pressure region. Again, we will arbitrarily define the range of pressures in the high-vacuum pressure regime as pressures from 1×10^{-3} torr to approximately 1×10^{-8} torr.

Over this range, the molecular density of gas molecules ranges from a high of 4×10^{13} molecules per cubic centimeter at one millitorr to a low of 4×10^{8} molecules per cubic centimeter at 1×10^{-8} torr. Over this same pressure range, the mean free path of the gas molecules increases from approximately 5.1 millimeters at one millitorr to 5 km at 1×10^{-8} torr.

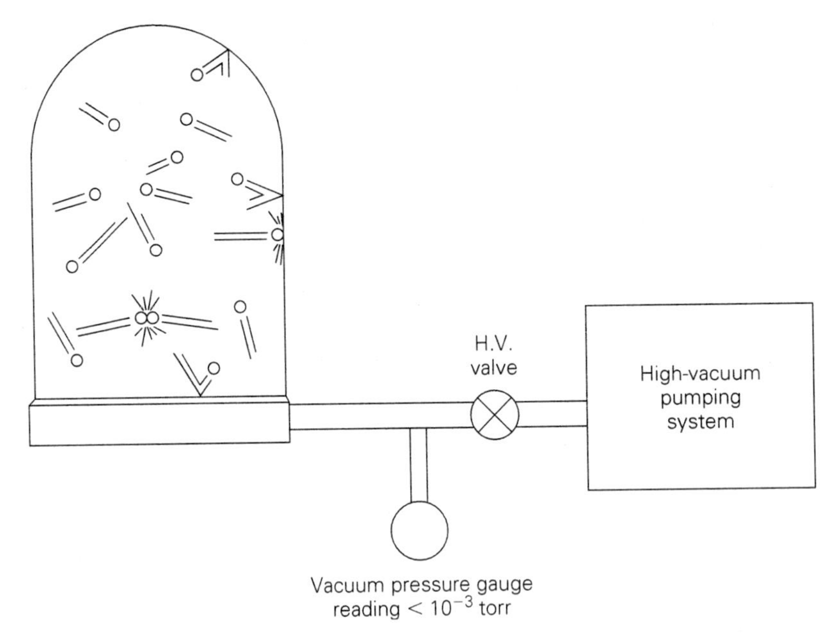

FIGURE 5.1
Molecular flow in the high-vacuum regime.

Under the rarefied conditions that we find in the chamber in the high-vacuum regime, gas molecules can easily travel distances greater than the dimensions of the process chamber before colliding with other gas molecules. Hence, the distance of travel in any single direction is limited only by the dimensions of the chamber, and collisions between gas molecules and the chamber walls will dominate. Flow in the high-vacuum regime is called *molecular flow*, and is illustrated in Figure 5.1.

5.2 GAS LOAD IN THE HIGH-VACUUM REGIME

The gas load is much more complex in the high-vacuum regime than in the rough vacuum regime because we are working with a much more rarefied environment (see Figure 5.2). Any small source of gas can add significantly to the gas load.

The high-vacuum pumps will first have to remove any bulk gas that the rough vacuum pumps were not able to remove. This is usually carried out very quickly.

If the chamber has been opened to the atmosphere, water molecules will be desorbing from the inner walls of the chamber. High-vacuum pumps can remove any water molecules from the chamber only as fast as they desorb from the surface and enter the gas phase. Outer layers of water molecules desorb much easier than the water molecules held closer to the surface of the chamber walls.

Real leaks can add significantly to gas load. Care in handling and installing vacuum components will ensure the integrity of your system. Virtual leaks are not real leaks, in that

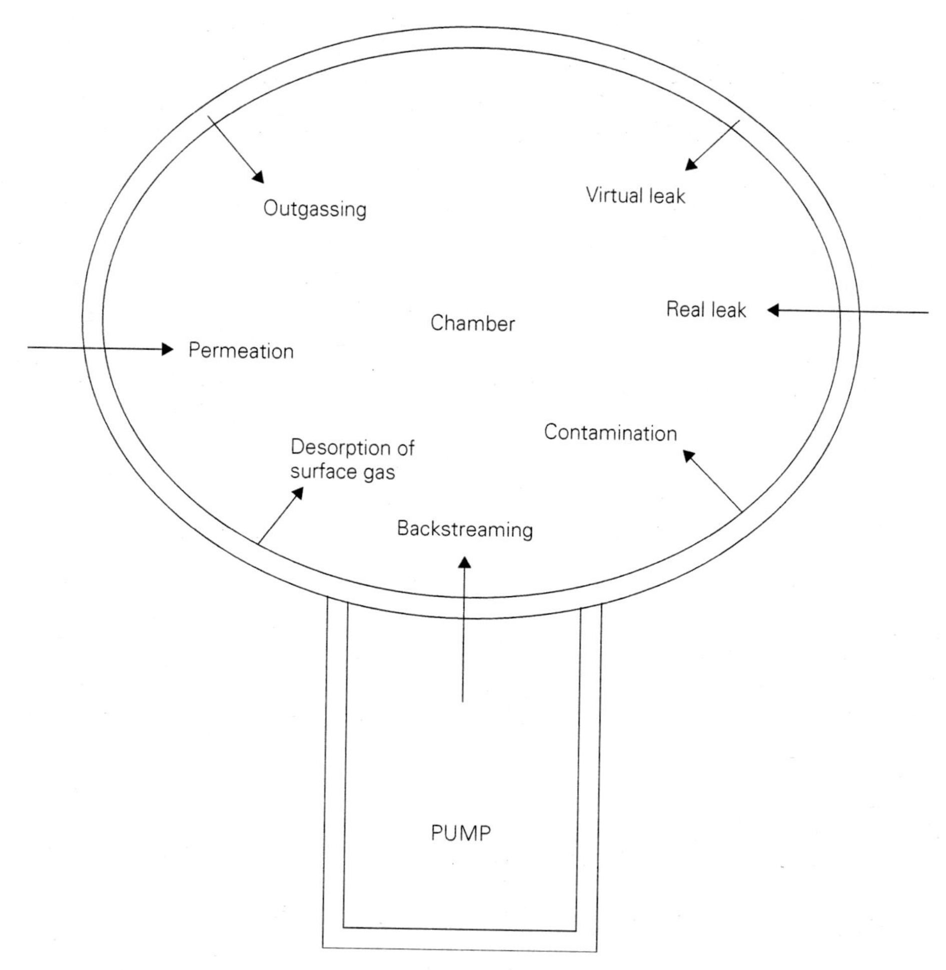

FIGURE 5.2
Gas load in the high-vacuum regime.

the gas does not come from outside the chamber. The term *virtual leak* refers to a trapped pocket of gas within the vacuum system that escapes into the chamber as the chamber is being pumped down. Gases can be trapped in welds, screws, and other places in the vacuum system. We will return to this topic in Chapter 7.

Gases can also permeate through materials such as elastomer O-rings; that is, they pass from outside the chamber through the O-ring materials, and add to the gas load in the chamber. The permeation rate of gases through metals is very low, but gas trapped in material like stainless steel during its manufacture may be emitted in the chamber. The process by which gases embedded in materials escape is called *outgassing*.

Contamination can also be a serious problem when you are trying to pump down a chamber into the high-vacuum regime. Contamination can occur when fingerprints, greases and oils, and other substances are left on the exposed surface of a vacuum chamber. These materials will change to a vapor at low pressures and add to the gas load.

Finally, *backstreaming* can be a problem if pumps using oil are used. Backstreaming occurs when oil vapors travel from the pump, against the flow of other gas molecules, back into the chamber. To eliminate this possibility, dry or oil-free pumps are used as backing pumps.

5.3 HIGH-VACUUM SYSTEMS

Figure 5.3 is a block diagram of a generic high-vacuum system. A high-vacuum system must have a combination of pumps that includes a rough vacuum pump and one or more high-vacuum pumps. A diaphragm backing a turbomolecular pump would work for small chambers. Larger chambers might require a rotary-vane backing a turbomolecular pump or a Roots pump.

There are two ways of roughing down the chamber. One method is rough the chamber through the high-vacuum pump, as shown in Figure 5.3. This can be accomplished if the high-vacuum pump is a turbomolecular pump. The second method is to use a separate roughing line that connects the rough vacuum pump directly to the chamber, by-passing the high-vacuum pump completely during the rough pump-down phase of operation.

The high-vacuum system, like the trainer shown in Figure 5.4, must have two types of vacuum gauges. One or more gauges will measure pressures in the rough vacuum regime

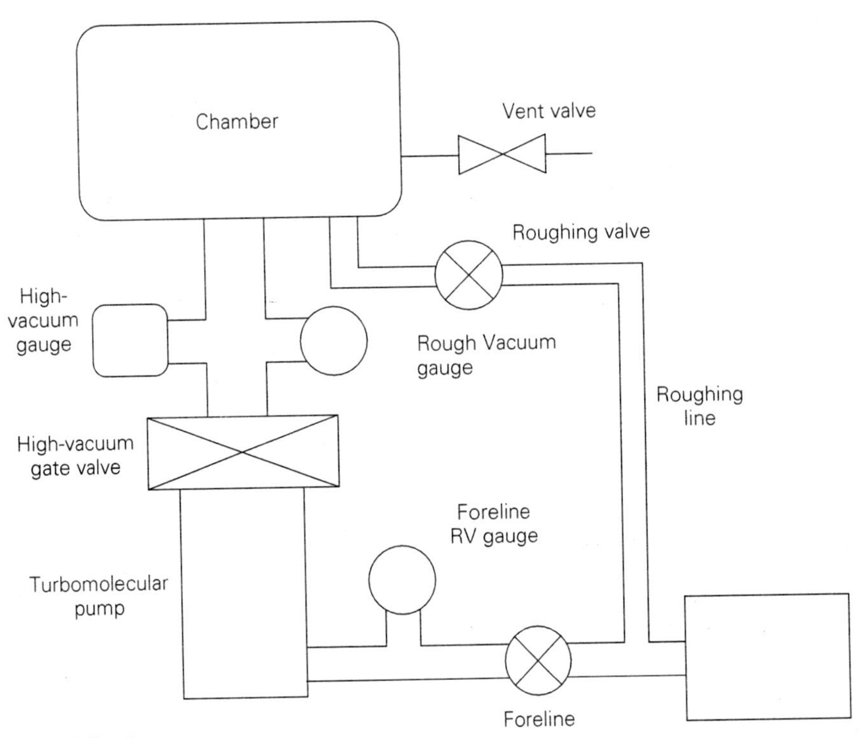

FIGURE 5.3
Simple high-vacuum system.

at various points in the system, such as the roughing line and the chamber. At least one high-vacuum gauge must be used, and its placement will most likely be such that it measures chamber pressure. All of these gauges will require a vacuum gauge controller to convert the gauge signals into pressure readouts so that the technician or operator can easily read the pressure.

A variety of valves can be strategically placed in the system to control gas flow. One valve will control gas flow through the roughing line. Another valve might control gas flow in the foreline of the high-vacuum pump, and a linear gate valve must be placed between the high-vacuum pump and the chamber. And finally, there must be one or more vent valves so that the system can be brought back to atmosphere.

Stainless steel will usually be the material of choice. Special types of stainless steel have low outgassing rates that are requisite to drawing a high-vacuum. Anodized aluminum can also used, but brass and other metals having high outgassing rates should be avoided.

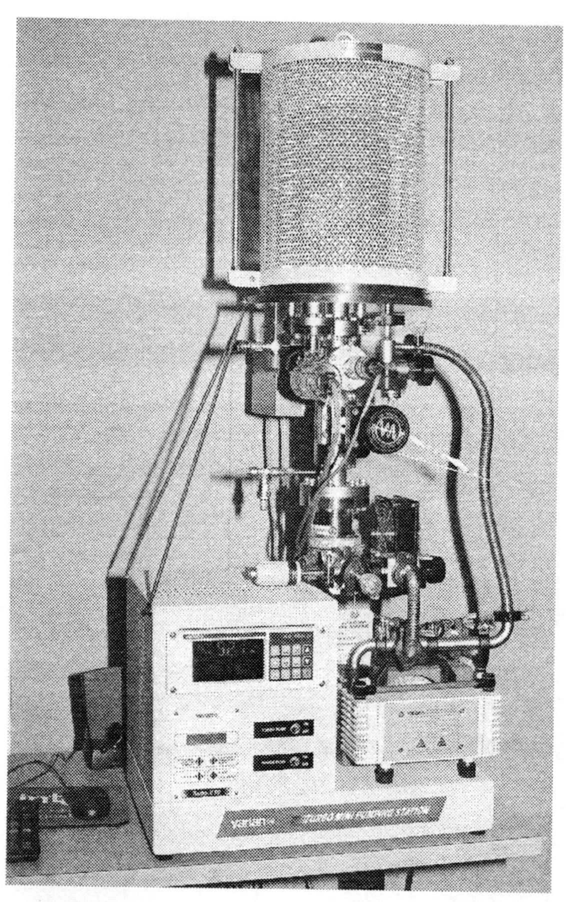

FIGURE 5.4
High-vacuum trainer.

5.4 HIGH-VACUUM PUMPS

High-vacuum pumps fall into two classifications: (1) kinetic vacuum pumps of the gas transfer type, and (2) entrapment pumps. *Turbomolecular pumps* fall into the kinetic vacuum pump classification. As with gas transfer rough vacuum pumps, gas molecules enter the inlet of the pump, are moved through the pump by the mechanical pumping mechanism, and are exhausted at the pump's outlet.

Entrapment pumps, on the other hand, retain the gas molecules on the inner surfaces of the pump. Two types of entrapment pumps are the cryopump and the sublimation pump. *Cryopumps,* as their name implies, use very cold surfaces to essentially "freeze" thc gas molecules, turning them to solids within the pump. As long as the pump is kept at its very cold operating temperature, the gases will remain in the pump. *Sublimation pumps,* on the other hand, use very reactive materials such as titanium to react with the gas molecules, converting them to solid by-products that stick to the inner surfaces of the pump. Once the gas molecules are frozen or changed to a sticky, solid material, they are essentially kept from returning to the chamber as gas atoms or molecules. Let us examine the pumping mechanism for each of these pumps, beginning with the turbomolecular pump.

5.4.1 Turbomolecular Pump

The turbomolecular pump, or turbo pump, is a kinetic-type high-vacuum pump. This implies that the mechanical action of the turbo pump imparts kinetic energy to the gas molecules and results in movement of gas molecules through the pump from inlet to outlet. The turbo pump accomplishes this pumping action through the interplay of slotted rotor and stator blades. The internal construction of a turbomolecular pump is shown in Figure 5.5.

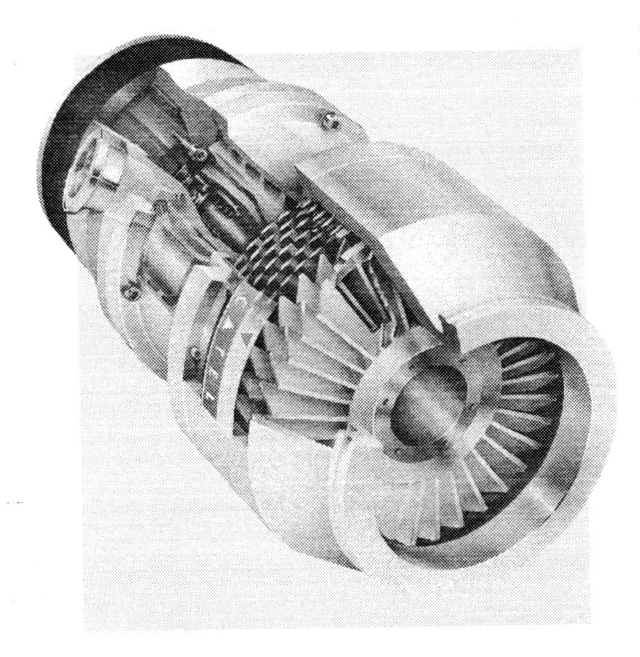

FIGURE 5.5
Turbomolecular pump.
ATS 200 with ceramic ball bearings. *Source:Brochure: Innovative Technology: Hybrid Turbomolecular Drag Vacuum Pump, ATS 100, ATS 200.* P/N 791816/MC, July 1995.

The turbo pump operates at rotor speeds that range from 80,000 rpm for small pumps to 24,000 rpm for large pumps. Each blade is only able to support a small pressure difference, so many blades are cascaded to obtain operating compression and pressure ratios.

The turbo pump works in the molecular flow region. Hence, the inlet of the turbo pump must be as large as possible to give the gas atoms or molecules the largest possible target and thus maximize the probability that the gas atom or molecule will enter the pump. The outlet of the turbo pump must also be in the molecular or transition flow regime. This requires that a roughing pump be used in series with the turbo pump to ensure that the foreline pressure at the exhaust of the turbo pump meets this condition. The ratio of the backing pressure to the inlet pressure is called the *compression ratio.*

The maximum compression ratio for a given gas occurs at a flow rate of zero. This is only a theoretical value, because there is always a gas flow, or *throughput,* into a high-vacuum pump due to the desorption rates of the installed parts on the high-vacuum side and also the diffusion of gas through the materials used to construct the vacuum chamber and associated components. The effective compression ratio of a turbomolecular pump is dependent on the gas throughput, and the compression ratio decreases as throughput increases.

As gas molecules enter the inlet of the turbomolecular pump, they encounter the rotating turbine blades. These angled blades hit the gas molecule, transferring the mechanical energy of the blades into gas molecule momentum. In this manner, gas molecules are directed from the inlet of the pump toward the exhaust port of the pump as gas is compressed from stage to stage (see Figure 5.6).

Let's take a look at the data sheet for a small turbomolecular pump, the Varian Turbo-V70LP. Table 5.1 lists the technical specifications for the Turbo-V70LP.

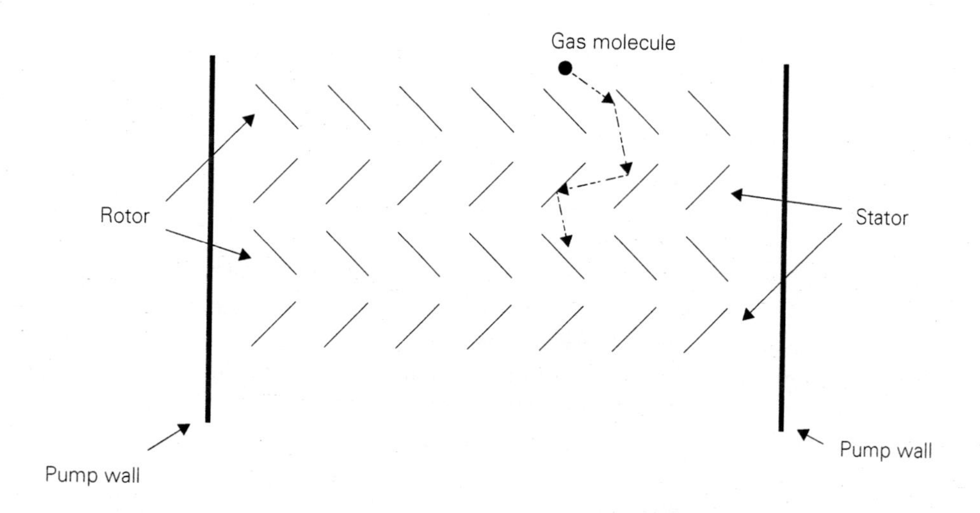

FIGURE 5.6
Momentum transfer to a gas molecule in a turbomolecular pump.

TABLE 5.1
Technical specifications for a Varian V70LP turbomolecular pump.

Technical specifications					
Pumping speed l/s	N2, 68	He, 60	H2, 45	**Recommended forepump**	Two-stage rotary pump: SD 40 Diaphragm pump:MDP 12
Compression ratio	N2, 5×10^8	He, 8×10^4	H2, 1×10^4		
Base pressure	With remomended mechanical pump			**Operating position**	Any
	2×10^{-10} mbar(1.5×10^{-10} torr) With recommended diaphragm pump: 2×10^{-8} mbar (1.5×10^{-9} torr)			**Cooling requirements**	Natural air convection (with low gas load) Forced air or water optional
				Bakeout temperature	120°c (CFF version)
Inlet flange	DN 63 ISO DN 40 KF	DN 63 CFF (4.5 in. OD) DN 35 CFF (2.75 in. OD)		**Vibration level (displacement)**	<0.01 μm at inlet flange
Foreline flange	NW 16 KB			**Weight lbs (kg)**	ISO: 3.7 (1.7); CFF: 5.5(2.5)
Rotational speed	75,000 rpm				
Start time	<1 min				

Source: Varian Vacuum Products Catalog, 2000, p. 151.

The first entry in the technical specifications is pumping speed. Three different pumping speeds are given: 68 L/s for nitrogen, 60 L/s for helium, and 45 L/s for hydrogen. This indicates that turbomolecular pumps, like many other types of pumps, are less efficient at pumping the lighter gas molecules. If we are pumping the chamber down from atmosphere, the pumping speed for nitrogen will be most useful (see Figure 5.7).

Compression ratios reflect the different pumping speeds for the three gases. The compression ratio for nitrogen is the highest at 5×10^8, whereas the compression ratios for helium and hydrogen are 8×10^4 and 1×10^4 respectively.

Base pressure, or ultimate pressure, is given for two scenarios. First, with the recommended mechanical pump, the SD-40 two-stage rotary pump, the base pressure is given as 1.5×10^{-10} torr. Second, with the recommended diaphragm pump, the MDP-12 two-stage diaphragm pump, the base pressure is given as 1.5×10^{-9} torr. These are optimum base pressures for best-case conditions, and therefore are not practical with any gas flow.

The V70LP can be ordered with ISO, KF, or CFF inlet flanges. This provides some flexibility in connecting the pump to the rest of the vacuum system. The outlet, or foreline, flange is a NW16 KF flange.

The rotational speed is 75,000 rpm. This is at the high end of rotational speeds for turbomolecular pumps. Larger turbomolecular pumps, because of the larger mass of the turbine blades, have much slower rotational speeds. The Turbo-V6000's rotational speed, for instance, is only 14,000 rpm.

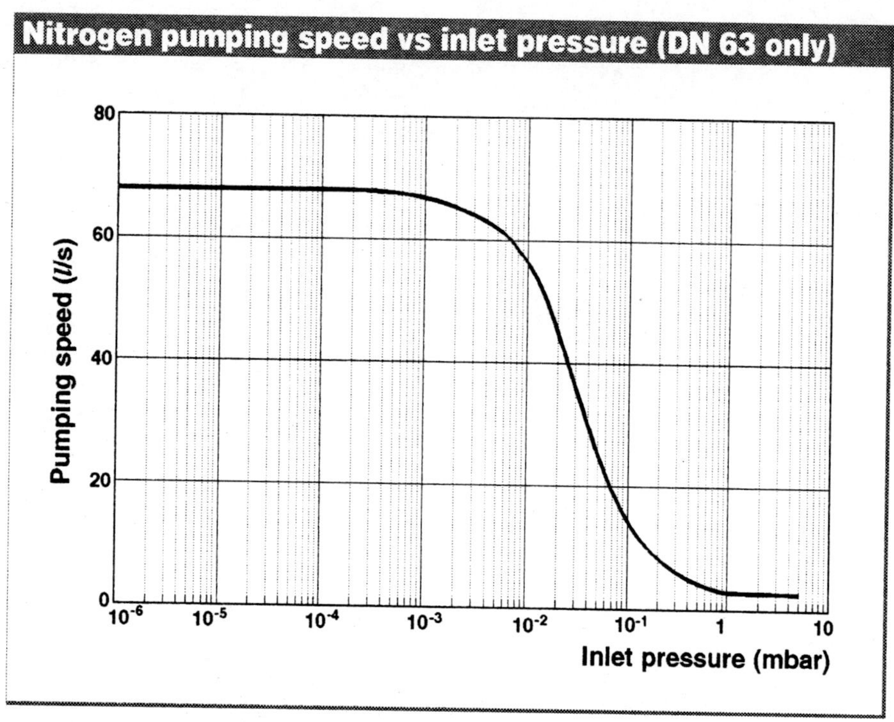

FIGURE 5.7
Pumping Curve for a VLP-70 turbomolecular pump.
Source: Varian Vacuum Products 2000 Catalog, p. 150.

Startup time for the V70LP is less than, or equal to, one minute. This is fairly fast and reflects the small size of the pump. The startup time for the larger V-6000 turbomolecular pump is approximately 30 minutes.

The V70LP and other small turbomolecular pumps can be operated vertically, horizontally, or in any other position. Other turbo pumps require a specific orientation for operation. For example, larger turbo pumps such as the V-6000 must be operated in the vertical position only.

Turbo pumps may require cooling. Cooling requirements for the V70LP are natural air convection for low gas loads, and forced air or water is optional. On the Varian Vacuum Trainer, a small fan is installed to provide forced-air cooling. On other turbo pumps, water or other types of cooling may be required.

Other pump specifications include bakeout temperature and vibration level. The V709LP with a CFF inlet flange can be baked out at a temperature of 120°C to remove water adhering to the interior surfaces of the pump. The vibration level or displacement is specified as less than 0.01 μm at the inlet flange. The larger V-6000 turbo pump has a vibration level of 0.05 μm or less.

The V70LP and other turbo pumps will require a controller. Controllers are microprocessor-controlled frequency converters with self-diagnostic and protection features.

The front panel display provides pump status, including rotational speed and error code diagnostics. It can also serve as a pump-cycle log and can display the number of vacuum cycles, the cycle time for the current cycle, and total operating hours on the pump. Remote operation can also be accomplished with logic level contact closures or the optional RS 232 line.

And finally, turbo pumps can come with a number of accessories. These may include inlet screens to block objects from entering into the pump, heater bands for bakeouts, water and air cooling kits, vibration dampers, and vent devices/flanges. All of these are available at extra cost, of course.

Turbomolecular pumps cost from several thousand dollars to tens of thousands of dollars. Add another few thousand dollars for the controller and additional dollars if you select any accessories. Hey, one never said that high-vacuum systems were inexpensive!

Selecting a turbomolecular pump depends on the application. If the application is to create a vacuum in a system where the gas load is produced only by outgassing, then the required pumping speed is given by

$$S_{eff} = \frac{Q}{p} \qquad \textbf{(5.1)}$$

where Q is the gas load due solely to outgassing at the desired pressure in $\frac{\text{torr} \cdot \text{L}}{\text{sec}}$, p is the desired base pressure in torr, and S_{eff} is the effective pumping speed.

EXAMPLE 5.1 Determine the effective pumping speed for a turbo pump and stainless steel chamber. The interior surface area of the chamber is 600 cm^2, the outgassing rate is $5 \times 10^{-8} \frac{\text{torr} \cdot \text{liters}}{\text{sec} \cdot \text{cm}^2}$, and the desired base pressure is 1×10^{-6} torr.

Solution

Using Equation 5.1

$$S_{eff} = \frac{5 \times 10^{-8} \frac{\text{torr} \cdot \text{liters}}{\text{sec} \cdot \text{cm}^2} \cdot 600\ \text{cm}^2}{1 \times 10^{-6}\ \text{torr}}$$

$$S_{eff} = 30\ \frac{\text{liters}}{\text{sec}}.$$

If the operation of the turbo pump involves the flow of process gases, then the effective pumping speed is a function of the process gas flow rate and the desired process pressure,

$$S_{eff} = \frac{Q_{process\ gas}}{p_{operating}} \qquad \textbf{(5.2)}$$

where $Q_{process\ gas}$ is the gas load due solely to the process gas flow at the desired pressure in $\frac{torr \cdot L}{sec}$,

$p_{operating}$ is the desired operating pressure in torr,

and S_{eff} is the effective pumping speed.

EXAMPLE 5.2 Determine the effective pumping speed required for a process that uses 20 sccms of process gas at a process pressure of 10 millitorr.

Solution

The first task is to convert 20 sccms of gas flow into a gas load in $\frac{torr \cdot L}{sec}$.

$$20\ sccm = \frac{20\frac{cc}{min} \cdot 760\ torr}{1000\frac{cc}{L} \cdot 60\frac{sec}{min}} = 0.253\frac{torr \cdot L}{sec}$$

Using Equation 5.2,

$$S_{eff} = \frac{0.253\frac{torr \cdot L}{sec}}{0.01\ torr}$$

$$S_{eff} = 25.3\frac{L}{sec}$$

5.4.2 Cryopump

The cryopump is a capture-type vacuum pump that operates by condensing and trapping gases on progressively colder surfaces inside the pump. It works like the freezer compartment of your refrigerator. Moisture-laden air enters the freezer compartment when the freezer door is opened. When the water vapor comes in contact with the cold coils in the freezer, it freezes onto the surface of the coil, forming a coating of ice. Since water vapor is being converted to solid ice, the partial pressure of water in the freezer drops. As a result, the total pressure decreases and creates a vacuum. This vacuum is short-lived, because the seal on the freezer door is not perfect and eventually the pressure increases back to atmospheric pressure.

We also learn from our freezer example that the capture process for water vapor has limits. As the ice gets thicker and thicker, the ability to capture more water molecules slows and the freezer functions less efficiently. Eventually, it is time to remove the ice, a process we call "defrosting" the freezer. Defrosting restores the efficiency of the freezer's coils.

The coils in a food freezer are approximately 0°F, cold enough to preserve food and keep the ice cream hard. On the other hand, the transformation of water vapor to ice is an unwanted by-product.

A cryopump operates by condensing and trapping gases on progressively colder surfaces. How cold do these surfaces have to be to condense not only water vapor but other gases, such as nitrogen, oxygen, argon, helium, and hydrogen? The graph shown in Figure 5.8 will

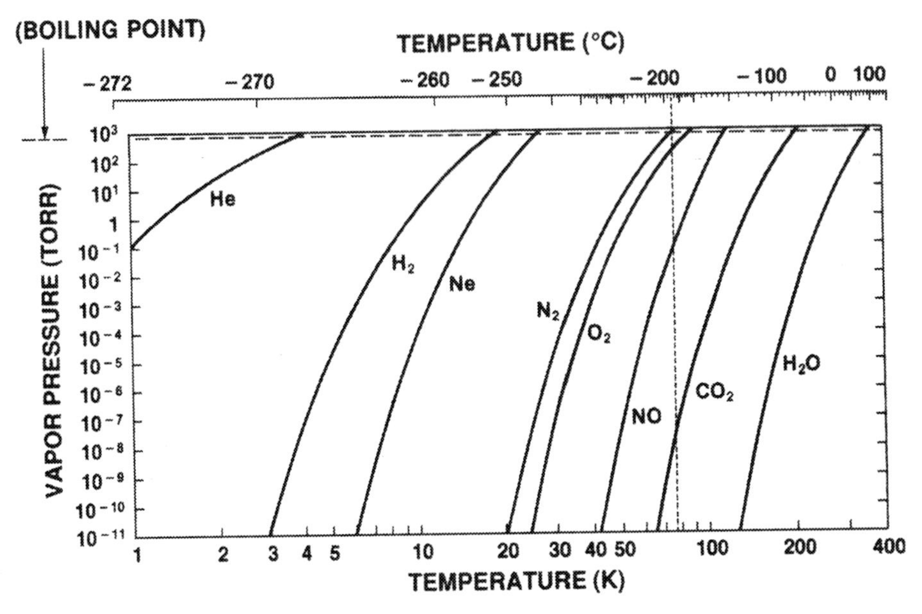

FIGURE 5.8
Vapor pressure versus temperatures curves for common gases.
Source: Varian Vacuum Products, *Basic Vacuum Practice,* p. 97.

give us this information. The graph shows the vapor pressure for common gases as a function of temperature.

In Figure 5.8, the warmer temperatures are on the right-hand side of the graph. The rightmost curve is for water. If the temperature is less than 125 Kelvin, water will have a very low vapor pressure and almost all of the water vapor will condense as ice on cold surfaces while the other gases remain in the gaseous state. If the surface temperature is lowered to 20 Kelvin, then additional gases will condense, including nitrogen, oxygen, carbon dioxide, leaving only the very light gases such as helium, hydrogen, neon in the gaseous state. Unfortunately, these very light gases require a temperature very near absolute zero to condense. Getting a surface this cold is not possible, so we have to look for other ways to trap these gases.

Figures 5.9a, b shows the internal structure of a cryopump. Near the inlet of the cryopump is the first cold surface that the gas molecules encounter. This inlet array, or *cryo panel,* operates at 60 to 100 K, and its job is to condense water vapor and carbon dioxide. Below the louvers is the cold head, which is cooled to a temperature of 10 K to 20 K. The cold head surface is cold enough to freeze argon, oxygen, and nitrogen.

Underneath the cold head are charcoal panels. These trap the lighter gas molecules by a process called *cryosorption.* In this process, gas molecules enter the porous charcoal and then cannot easily find their way out.

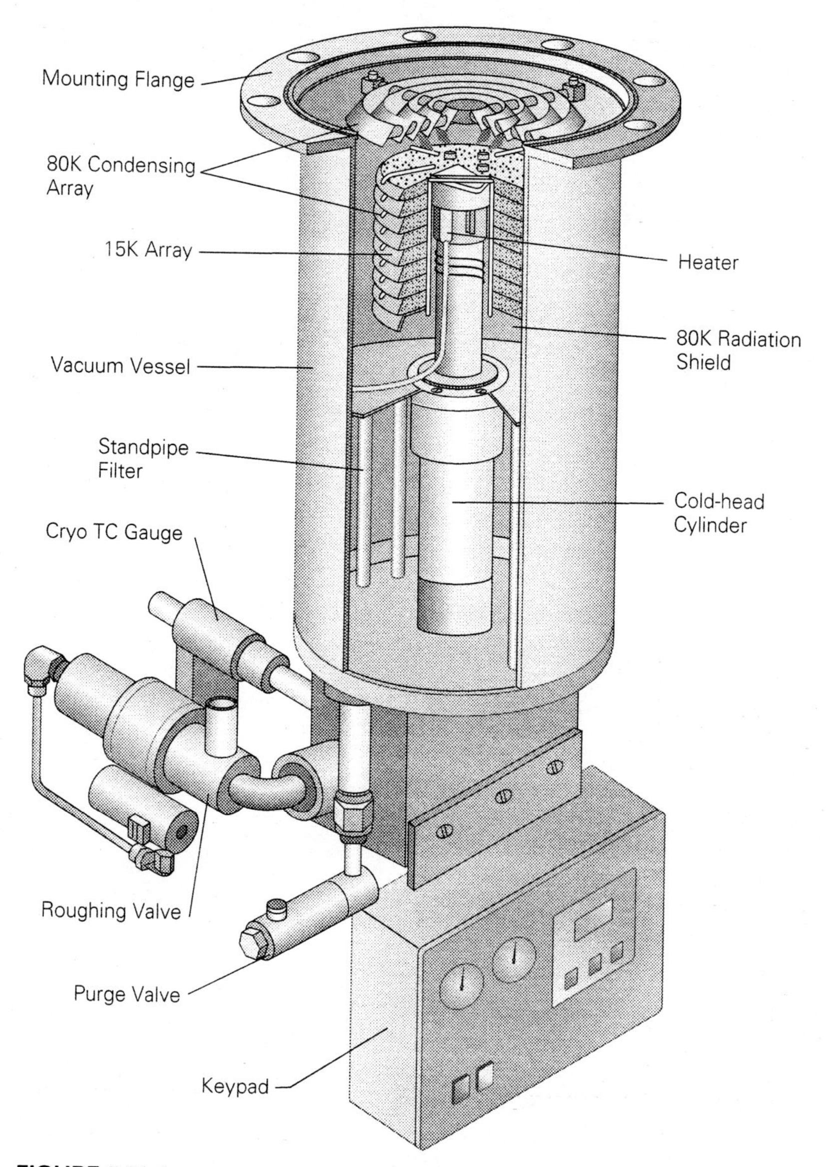

FIGURE 5.9(a)
Cross-section of a cryopump.
Source: Gary Ash, "Cryogenic High Vacuum Pumps," *Vacuum & Thinfilm,* August 1999, p. 20.

In some applications, a cooled radiation shield is placed at the inlet of the cryopump. The function of this cooled radiation shield is to minimize the effect of radiation or hot gases emanating from the process chamber. Inadequate shielding from this radiation and/or hot gases will cause the cryopump to release a pressure pulse when the process is started, because some of the trapped gas will have warmed sufficiently to escape from the pump.

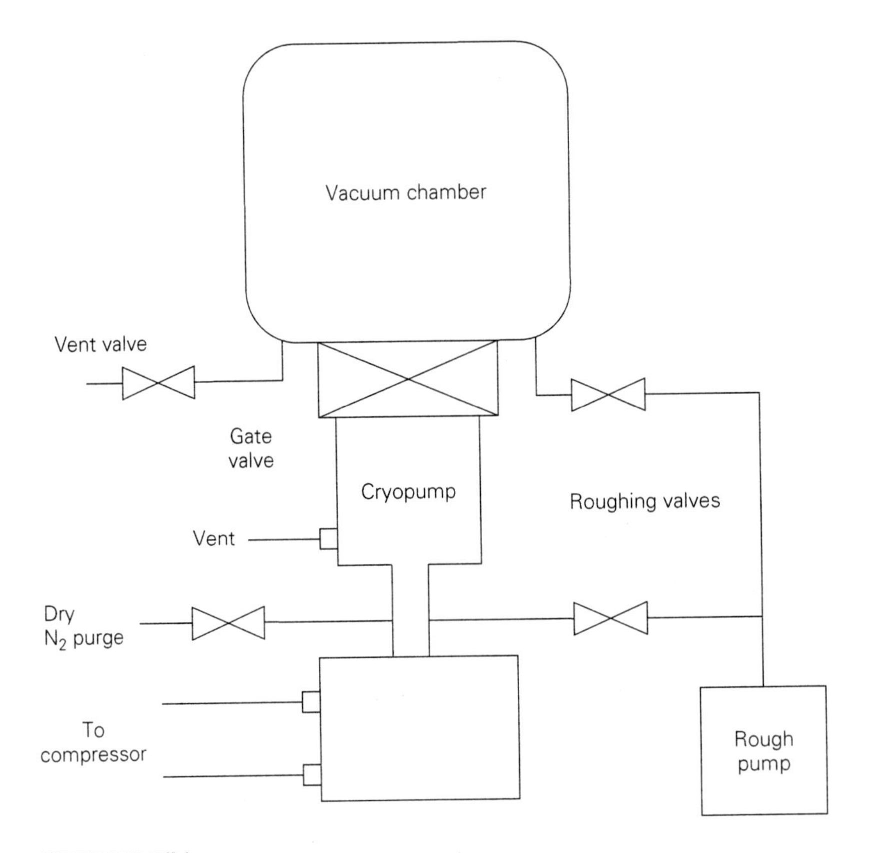

FIGURE 5.9(b)

This sounds simple. Use a refrigeration system to make some very cold surfaces, and use these cold surfaces to freeze the heavier gas molecules that flow into the pump. Then use charcoal to cryosorb the lighter, more energetic gas molecules. But it is not quite so simple.

At room temperature, the speed of a molecule is a function only of its molecular mass. The lighter molecules have the highest speed, and the heavier molecules have the slowest speed. For example, hydrogen, with a molecular mass of 2, has a speed of 44.6 liters/second through an opening with an area of 1 square centimeter. The heavier water molecule, mass of 18, has an ideal speed of 14.9 liters per second, and nitrogen, which is heavier yet, mass of 28, has an ideal speed of 11.9 liters per second. Based on these speeds, one might think that hydrogen would be pumped the fastest and nitrogen the slowest, but we know that cryopumps are most efficient at pumping water. How can this be?

If all the gas molecules that hit the face area of the cryopump were to freeze, then these ideal speeds would be the pumping speed of the pump for these gases. Water approaches the ideal value, indicating that almost all the water molecules that hit the inlet cryopanels or louvers stick to the cold surface without rebounding back into the chamber. Thus, the pumping speed of a cryopump is approximately 14.9 liters per second per square centimeter.

Gases such as nitrogen, oxygen, and argon have to pass through the louvers to reach the inner cold panel. Only a fraction of the incoming molecules for these gases reach the cold panel. The others bounce off the inlet louvers and rebound back into the chamber. A cryopump that has an inlet louver allows only about 25% of the air molecules (nitrogen and oxygen) to freeze out on the cold panel. For nitrogen, the net pumping speed is only 25% of the ideal pumping speed of 11.9 liters per second per square centimeter, or about 3.0 liters/second/cm^2.

The lightest gas molecules (hydrogen and helium) have a more arduous path to travel to reach the charcoal layer under the cold panel. As a result, only about 12% of the very light gas molecules that hit the face of the cryopump are actually cryosorbed. Thus the net pumping speed for hydrogen and helium is only about 12% of 44 liters/second/cm^2, or 5.6 liters/second/cm^2.

Regeneration As condensed gas builds up inside a cryopump, the pump gradually loses its pumping capacity. When the pumping speed drops below an acceptable level, the pump must be restored to its original condition through a process known as *regeneration*. A pump is usually regenerated by allowing its cold surfaces to return to room temperature while it is purged with dry, heated nitrogen gas. Regeneration can take anywhere from three to eight hours. Due to these long regeneration times, some manufacturers offer the option of a partial regeneration process. These partial regens warm up the cold head for a shorter time, allowing process gases such as argon and hydrogen to blow off quickly, while not melting the ice on the cryo array. In this manner, several partial regens can be performed before a full regeneration procedure is needed.

5.5 HIGH-VACUUM GAUGES

Since the density of gas molecules is orders of magnitude lower in the high-vacuum regime than in the rough vacuum regime, the way we measure the pressure in the vacuum system must change. Rather than using the force exerted by gas molecules on a surface, as in a capacitance manometer, or the thermal properties of the gas, as in thermocouple and Pirani gauges, high-vacuum gauges use energetic electrons to ionize some of the gas molecules. The positively charged ions that are created are collected, and the current is measured by an electrometer in the gauge controller. Both the Bayard-Alpert gauge and the cold cathode gauge use this method to measure pressures in the high-vacuum regime.

5.5.1 Bayard-Alpert Ionization Gauge

The Bayard-Alpert gauge is an ionization gauge. That is, gas molecules are ionized by high-energy electrons, and the ions formed are collected by a grounded collector wire oriented along the axis of the cylindrical grid.

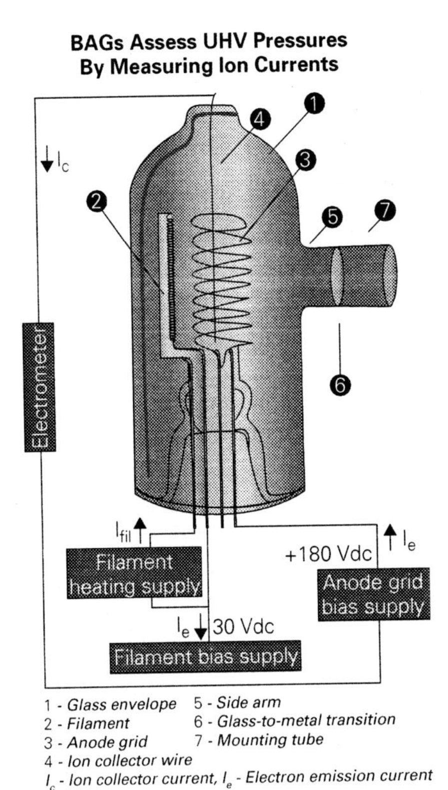

FIGURE 5.10
Bayard-Alpert gauge.
Source: R & D Magazine, February 2000, p. 69.

The structure of a glass-enclosed Bayard-Alpert gauge is shown in Figure 5.10. The outer enclosure is made of glass (Pyrex) and has a side arm attached to a vacuum fitting (KF or CFF) used to attach the gauge to the vacuum system. Inside the outer enclosure are the filament, grid, and ion collector.

The Bayard-Alpert gauge operates in the following manner. Electrons boiled off a hot filament are accelerated toward the negatively charged grid. Since the grid is fairly open, most of the electrons miss the grid wires and enter the cylindrical space within the grid. In this region, the energized electrons collide with gas molecules, and some of the gas molecules are ionized. The ions that are formed are then attracted to a negatively charged collector located along the axis of the cylindrical grid. The ion collector is connected to the gauge controller's electrometer, and the ion current is measured. If the electron emission current from the filament and the temperature of the gas are held constant, then the ion current is proportional to the pressure of the gas.

The operating range for a Bayard-Alpert gauge is from 10^{-3} torr to a lower limit in the 10^{-9} to 10^{-10} torr range with a suitable controller. The upper pressure limit is defined as the pressure where the ion current versus pressure relationship deviates from linearity. The lower limit is determined by an effect known as the *x-ray limit* caused by x-rays produced when highly energetic electrons collide with the grid structure. These x-rays can be intercepted by the ion collector and produce a current indistinguishable from the ion current produced by the ionization of gas molecules by electron bombardment.

The accuracy of a Bayard-Alpert gauge is ± 20%. Long-term stability is highly dependent on gauge construction, filament material, operating conditions, and the vacuum environment. The repeatability of glass-enclosed gauges is affected by the accumulated electrostatic charge on the glass wall due to electrons from the filament. Some manufacturers coat the inside the glass enclosure with a metal film to improve the repeatability.

The filaments in the Bayard-Alpert gauge are not replaceable, so the gauge is useless once the filament burns out. For this reason, Bayard-Alpert gauges are available in dual-filament designs. Filament materials include tungsten and thoria-coated iridium.

Glass-enclosed gauges may also collect gas molecules through a pumping action caused by chemical and electrical effects. To minimize this pumping action, a large conductance connection between the gauge and the rest of the vacuum system must be provided when making pressure measurements at the low end of the operating range. Glass when heated also increases the permeation of helium from the atmosphere and may be an issue when checking the vacuum system for helium leaks. To eliminate issues related to the glass enclosure, Bayard-Alpert gauges are manufactured as an all-metal, or "nude," gauge.

In the nude version of the Bayard-Alpert gauge (see Figure 5.11), the electrodes are not enclosed in glass. Instead, the electrodes are welded onto an insulating feedthrough mounted on a vacuum-compatible flange directly inserted into the vacuum environment. This allows gas molecules to freely flow into the ionization volume of the gauge and also eliminates the pressure differential normally associated with glass-enclosed gauges. Otherwise, the two types of gauges are identical or very similar.

Since the amount of energy needed to ionize different gas molecules differs, the Bayard-Alpert gauge is sensitive to gas type. The gauge is normally calibrated for nitrogen. A table of relative sensitivities like the one shown in Table 5.2 provides a means of correcting the gauge reading for other gases.

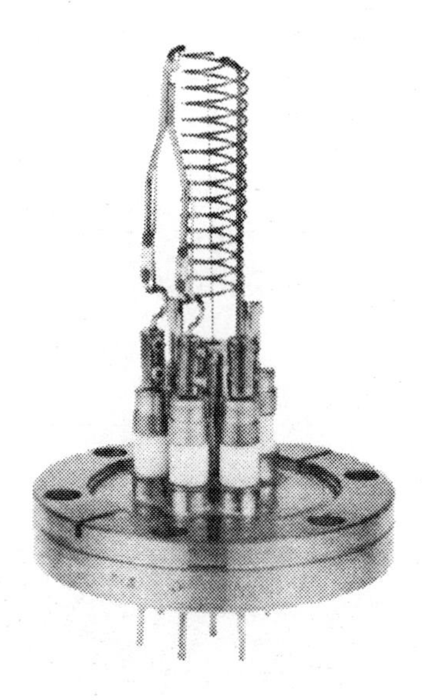

FIGURE 5.11
Nude Bayard-Alpert gauge.
Source: Varian Vacuum Products Catalog 2000, p. 317.

TABLE 5.2
Relative gas sensitivities for a Bayard-Alpert gauge.

Gas	Nominal gas correction factor
Helium, He	0.18
Neon, Ne	0.30
Hydrogen, H_2	0.46
Nitrogen, N_2	1.00
Air	1.0
Oxygen, O_2	1.01
Carbon monoxide, CO	1.05
Water vapor, H_2O	1.12
Nitrous oxide, NO	1.15
Ammonia, NH_3	1.23
Carbon dioxide, CO_2	1.42
Methane, CH_4	1.4
Krypton, Kr	1.94
Sulfur hexafluoride, SF_6	2.2
Ethane, C_2H_6	2.6
Xenon, Xe	2.87
Mercury, Hg	3.64
Propane, C_3H_8	4.2

5.5.2 Cold Cathode Gauge

The cold cathode gauge, shown in Figure 5.12, is an alternative to the Bayard-Alpert gauge in the high-vacuum regime. Like the Bayard-Alpert gauge, the cold cathode gauge, or Penning vacuum gauge, ionizes gas molecules by electron impact. The cold cathode gauge does this in a high-voltage discharge. An external, permanent magnet causes the electrons to travel in helical paths, thereby increasing the probability that gas molecules will be ionized. The positive ions that are formed are drawn to the negatively charged cathode. Like the Bayard-Alpert gauge, the pressure-dependent current is measured by an electrometer in the gauge controller.

Conventional Penning-type sensors do not work well below 10^{-6} torr because of the difficulty in maintaining the discharge at low pressures. Modern cold cathode gauges use an inverted magnetron design that has an isolated ion collector, making the sensor less susceptible to contamination and allowing a wider range of pressure measurement, into the 10^{-9} torr pressure range.

The operation of an inverted magnetron gauge can be described in the following manner. A high voltage of several kilovolts is applied between the anode and the cathode. After a few seconds at pressures above 10^{-6} torr, a glow discharge is ignited in the gauge. The electric and magnetic fields cause the electrons in the glow discharge to travel in helical paths, increasing the chance of collisions with gas molecules. The ions

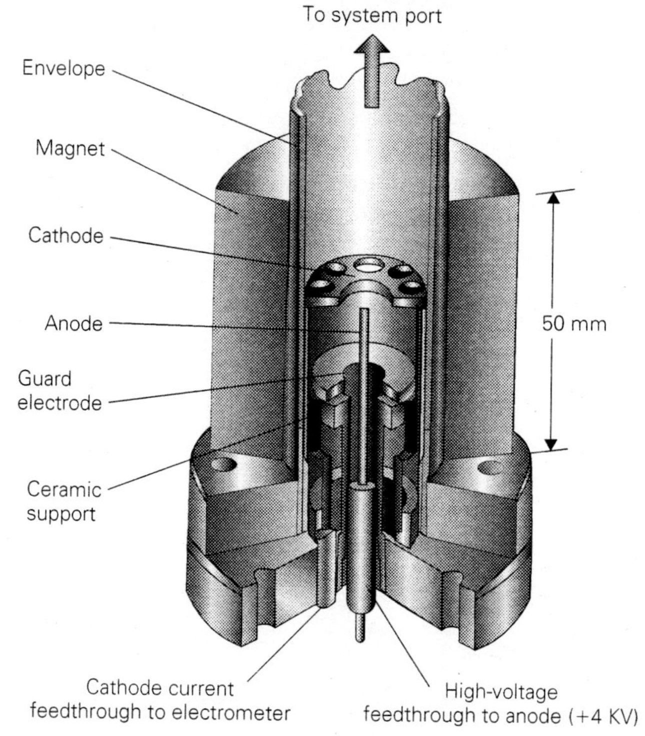

FIGURE 5.12
An Inverted magnetron high-vacuum gauge.
Source: R & D Magazine, October 2000, p. 35.

are collected by the cathode and result in a sensor current that is proportional to the pressure.

The advantages of cold cathode gauges over Bayard-Alpert gauges include the following:

- There is no filament to break or burn out, making the cold cathode gauge immune to in-rushes of air and to damage due to vibration.
- Relatively immune to gases, such as oxygen and halogens, that would destroy a hot filament.
- Can be cleaned and reused almost indefinitely.
- Operates at ambient temperature, and thus does not require degassing.
- Lower pressure measurements are not masked by thermal outgassing of the sensor.
- No x-ray limit to lower pressure measurements.
- Control circuit has only one current loop, as opposed to three for a conventional Bayard-Alpert gauge.

5.6 CONDUCTANCE

Although gas flow in the high-vacuum regime is described as molecular flow, the movement of gas is not really a flow, but rather the random motion of gas molecules. This random motion of gas molecules results in a flow that occurs when gas molecules move from a region with a higher density of gas molecules to a region with a lower density of gas molecules over time.

Since the flow is now molecular flow rather than viscous flow, the equations for conductance are different. As shown in Figure 5.13, for a straight tube operating in the molecular flow region, the conductance of the aperture is given by

$$C_{aperture} = 11.6\frac{\mathrm{L}}{\mathrm{sec}\cdot\mathrm{cm}^2}A$$

$$\text{where} \quad A = \pi\frac{d^2}{4} \tag{5.3}$$

and the conductance of the tube is given by

$$C_{tube} = 12\frac{\mathrm{L}}{\mathrm{sec}\cdot\mathrm{cm}^2}\frac{d^3}{L} \tag{5.4}$$

where d is the diameter, and L is the length.

From the two conductance equations for a straight tube operating in the molecular flow regime, it becomes apparent that the bigger the diameter of the aperture, the greater the conductance, and the shorter the tube, the greater the conductance. Therefore, to maximize the conductance between the chamber and the high-vacuum pump, the pump should be attached directly to the chamber, and the aperture should be the same size as the inlet port of the pump.

Let's consider some examples and see how the conductance equations can be used to determine the conductance of a straight length of pipe.

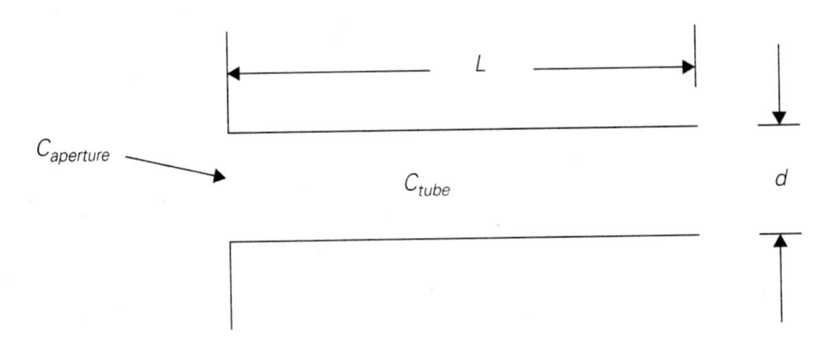

FIGURE 5.13
Conductance of a straight pipe.

EXAMPLE 5.3 Determine the net conductance of a tube 20 cm long and 2 cm in diameter.

Solution

Using the pipe dimensions given, the conductance of the aperture and the conductance of the tube can be found using Equations 5.3 and 5.4.

$$C_{aperture} = 11.6\frac{\text{L}}{\text{sec} \cdot \text{cm}^2}A$$

$$C_{aperture} = 11.6\frac{\text{L}}{\text{sec} \cdot \text{cm}^2}\frac{\pi(2\text{ cm})^2}{4}$$

$$C_{aperture} \approx 36.4\frac{\text{L}}{\text{sec}}$$

and

$$C_{tube} = 12\frac{\text{L}}{\text{sec} \cdot \text{cm}^2}\frac{d^3}{L}$$

$$C_{tube} = 12\frac{\text{L}}{\text{sec} \cdot \text{cm}^2}\frac{(2\text{ cm})^3}{20\text{ cm}}$$

$$C_{tube} \approx 2.4\frac{\text{L}}{\text{sec}}$$

The conductance of the aperture and the conductance of the tube can now be combined to find the net conductance.

$$C_{net} = \frac{C_{aperture} \cdot C_{tube}}{C_{aperture} + C_{tube}}$$

$$C_{net} = \frac{\left(36.4\frac{\text{L}}{\text{sec}}\right) \cdot \left(2.4\frac{\text{L}}{\text{sec}}\right)}{\left(36.4\frac{\text{L}}{\text{sec}}\right) + \left(2.4\frac{\text{L}}{\text{sec}}\right)}$$

$$C_{net} \approx 2.25\frac{\text{L}}{\text{sec}}$$

EXAMPLE 5.4 Determine the net conductance of a tube 20 cm long and 8 cm in diameter.

Solution

Using the pipe dimensions given, the conductance of the aperture and the conductance of the tube can be found using Equations 5.3 and 5.4.

$$C_{aperture} = 11.6 \frac{\text{L}}{\text{sec} \cdot \text{cm}^2} A$$

$$C_{aperture} = 11.6 \frac{\text{L}}{\text{sec} \cdot \text{cm}^2} \frac{\pi (8 \text{ cm})^2}{4}$$

$$C_{aperture} \approx 583 \frac{\text{L}}{\text{sec}}$$

and

$$C_{tube} = 12 \frac{\text{L}}{\text{sec} \cdot \text{cm}^2} \frac{d^3}{L}$$

$$C_{tube} = 12 \frac{\text{L}}{\text{sec} \cdot \text{cm}^2} \frac{(2 \text{ cm})^3}{20 \text{ cm}}$$

$$C_{tube} \approx 307 \frac{\text{L}}{\text{sec}}$$

The conductance of the aperture and the conductance of the tube can be combined to find the net conductance.

$$C_{net} = \frac{C_{aperture} \cdot C_{tube}}{C_{aperture} + C_{tube}}$$

$$C_{net} = \frac{\left(583 \frac{\text{L}}{\text{sec}}\right) \cdot \left(307 \frac{\text{L}}{\text{sec}}\right)}{\left(583 \frac{\text{L}}{\text{sec}}\right) + \left(307 \frac{\text{L}}{\text{sec}}\right)}$$

$$C_{net} \approx 201 \frac{\text{L}}{\text{sec}}$$

5.7 HIGH-VACUUM PUMP DOWNS

Performing a high-vacuum pump down from atmosphere is a three-step process. The first step is the rough pump-down phase, where the pressure is reduced from atmospheric pressure to a cross-over pressure to the high-vacuum phase. The second step is the cross-over from pumping with the rough vacuum pump to pumping with the high-vacuum pump. The third and final phase is the high-vacuum pump down to an ultimate pressure or working pressure. A typical pump-down curve is shown in Figure 5.14.

The rough vacuum pump down was discussed in chapter 4. To review, during this period of time, the rough pump is removing the volume or bulk gas in the chamber, and the chamber pressure falls from atmosphere toward the ultimate pressure of the pump. Factors affecting the pump-down time include the net pumping speed of the rough vacuum pump and piping, the surface condition, and water desorption. The pump-down time can be estimated using Equation 4.1,

$$t = \frac{V}{S_{eff}} \ln \frac{P_i - P_{ult}}{P_f - P_{ult}}.$$

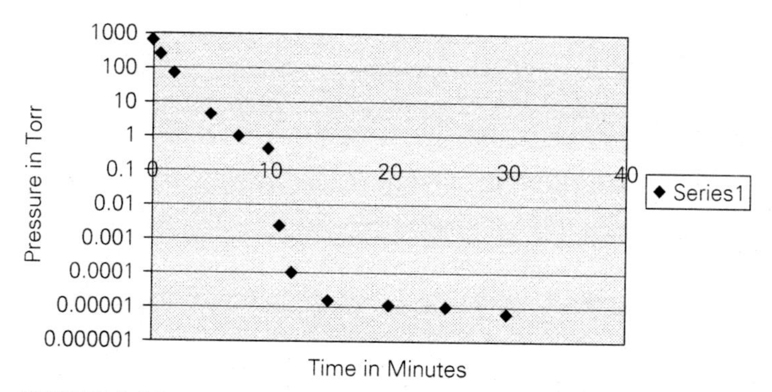

FIGURE 5.14
Typical high-vacuum pump-down curve.

Cross-over is usually defined as the point in time when the roughing valve is closed and the high-vacuum valve is opened, assuming that the system has separate roughing and high-vacuum lines. Sometimes the system is configured in such a manner that roughing occurs through the high-vacuum line. An example is a diaphragm pump roughing through a turbomolecular pump; in this case the cross-over would be defined as the point in time when the turbomolecular pump is turned on.

Immediately after the high-vacuum valve is opened to initiate cross-over, the pressure drops rapidly because of an increase in pumping speed. The last of the volume or bulk gas is quickly removed, and surface gas evolution, mainly water vapor, becomes the predominant gas load.

The final phase is the high-vacuum pump down. Estimating the pump-down time from the cross-over pressure to the ultimate pressure in the high-vacuum regime is difficult. In the literature, the high-vacuum pump-down time is described mathematically as

$$t = \beta\tau \ln \frac{p_c}{p - p_u}$$

where β is a correction factor, τ is the time constant of the outgassing process, p_c is the pressure immediately after switching to the high-vacuum pump, and p_u is the ultimate pressure due to the limitation of the high-vacuum pump and other constant gas load factors. The difficulty of getting accurate values for β, τ, and p_c makes it difficult to calculate high-vacuum pump-down times. On the other hand, the mathematical relationship tells us that the pump-down time is proportional to the time constant of the outgassing process. Since water vapor desorption is often a major component and has a relatively long time constant, the pump-down time to a pressure near the ultimate pressure will take a considerable amount of time.

Another computation provides an estimate of the ultimate pressure that can be achieved in a high-vacuum system. Ultimate pressure can be simply stated as follows,

$$P_{ultimate} = \frac{gas\ load}{net\ pumping\ speed}. \tag{5.5}$$

The net pumping speed can be calculated from the pumping speed of the high-vacuum pump and the conductance of the piping connecting the pump to the chamber. The gas load is a harder parameter to quantify. One can assume that the ultimate pressure occurs after almost all the water has desorbed from the chamber surface. If this is the case, we can look to other contributors to gas load, such as outgassing from the materials used to manufacture the chamber and exposed to the chamber and permeation through materials into the chamber. The following example gives an example of how ultimate pressure can be obtained in a high-vacuum system.

EXAMPLE 5.5

A cylindrical stainless steel tank is evacuated using a turbomolecular pump backed by a scroll pump. The tank is 30 cm in diameter and 50 cm in height. Assume that the outgassing rate for the stainless steel used to manufacture the tank is 5×10^{-8} torr liters/sec cm2 and the pumping speed of the turbo pump is 150 liters/sec. Also assume that water desorption is not a significant factor. Estimate the ultimate pressure that can be achieved.

Solution

To use Equation 5.5, we need to determine the gas load based on the outgassing rate of the stainless steel. We will begin by calculating the interior surface area of the chamber. The area of the two ends of the chamber can be found as follows:

$$Area_{ends\ of\ chamber} = 2\pi r^2$$

$$Area_{ends\ of\ chamber} = 2\pi\left(\frac{30\text{ cm}}{2}\right)^2$$

$$Area_{ends\ of\ chamber} \approx 1{,}414\text{ cm}^2$$

The surface area of the body of the chamber can be found as follows:

$$Area_{body\ of\ chamber} = 2\pi rh$$

$$Area_{body\ of\ chamber} = 2\pi\left(\frac{30\text{ cm}}{2}\right)(50\text{ cm})$$

$$Area_{body\ of\ chamber} \approx 4{,}712\text{ cm}^2$$

Adding the two surface area values together yields the total interior surface area of the stainless steel chamber.

$$Area_{Total} = Area_{ends\ of\ chamber} + Area_{body\ of\ chamber}$$

$$Area_{Total} = 1{,}414\text{ cm}^2 + 4{,}712\text{ cm}^2$$
$$Area_{Total} = 6{,}126\text{ cm}^2$$

Using the total area and the given outgassing rate for stainless steel, we can calculate the total gas load:

$$Gas\ load = area \times outgassing\ rate$$

$$Gas\ Load = 6{,}126\ \text{cm}^2 \cdot 5 \times 10^{-8}\ \frac{\text{torr} \cdot \text{liters}}{\text{sec} \cdot \text{cm}^2}$$

$$Gas\ Load \approx 3.06 \times 10^{-4}\ \frac{\text{torr} \cdot \text{liters}}{\text{sec}}$$

Finally, using the total gas load and pumping speed for the turbomolecular pump, we can estimate the ultimate pressure that we can achieve.

$$P_{ultimate} = \frac{3.06 \times 10^{-4}\ \dfrac{\text{torr} \cdot \text{liters}}{\text{sec}}}{150\dfrac{\text{liters}}{\text{sec}}}$$

$$P_{ultimate} \approx 2 \times 10^{-6}\ \text{torr}$$

5.8 WATER IN HIGH-VACUUM SYSTEMS

If a vacuum system has been exposed to air for any length of time, its surfaces will be coated with water. How much water vapor is there in air? Well, that depends on the temperature and humidity. Air at 25°C and 50% relative humidity contains approximately 12 torr of water vapor and is the third most prevalent gas. If the air is more humid, then the partial pressure of water vapor increases, and conversely, if the air is less humid, the partial pressure of water vapor decreases.

The nonporous materials of the vacuum system will adsorb water vapor in molecular monolayers. The first layers will adhere tightly to the surface atoms. Subsequent layers of water will adhere to existing monolayers of water molecules in ever-weakening bonds as the bed of sorbed water molecules becomes thicker. The thickness of the bed will be determined by the number of water molecules that impact the surface and the time of exposure. To minimize the adsorption of water on surface in a vacuum system, keep the time of exposure to air to a minimum.

A practical manufacturing system requires that fresh material be introduced into the process/vacuum chamber from the atmosphere and will transfer any water adhering to this material into the chamber. This applies to both batch and load-lock systems. To minimize the amount of adsorbed water, materials can be loaded into sealed storage containers directly from the load locks on the vacuum chamber.

The elastomer O-rings used to seal the flanges on vacuum systems are also a source of water vapor gas loads. When manufactured, elastomer O-rings are heavily loaded with water, and this water will outgas when placed in a vacuum. Prebaking O-rings can reduce the initially high outgassing rate. In addition, water vapor can permeate through elastomer O-rings. In high-vacuum systems, where low ultimate pressures are desired, elastomer O-rings are replaced with copper gaskets to eliminate these outgassing and permeation problems.

SUMMARY

Achieving pressures in the high-vacuum regime requires more work and more complex pumping systems. The gas load in this pressure regime is a combination of water desorption from the interior surfaces of the vacuum system, outgassing from the materials used to construct the system, and permeation through materials, as well as from unwanted sources such as leaks and contamination.

Usually a combination of pumps is required to achieve pressures in the high-vacuum regime. Two commonly used oil-free high-vacuum pumps are the turbomolecular pump and the cryopump. The turbomolecular pump uses high-speed turbine blades to kinetically move gas molecules from the process chamber, while the cryopump uses very cold surfaces to cause gas molecules to stick to the interior surface of the pump.

The gauges used for pressure measurements in the high-vacuum regime differ from those used in the rough vacuum regime. Common high-vacuum pressure gauges include the Bayard-Alpert gauge and the cold cathode gauge.

The geometry of the vacuum subsystem plays a much greater role in determining pump-down times. The conductance of the piping connecting the pumps to the process chamber may have a limiting effect on the net pumping speed and thus the overall pump-down time.

Removing water from high-vacuum systems is a major problem. Desorption of water from interior surfaces may be speeded up by purging the process chamber with dry nitrogen gas or by internally heating the process chamber using infrared lamps.

BIBLIOGRAPHY

"Advanced Vacuum Practices." Course Workbook, Varian Vacuum Products.

Hablanian, Marsbed H., *High-Vacuum Technology: A Practical Guide,* 2nd Edition, Marcel Dekker, New York, NY, 1997.

Bopp, Eric. "The Advantages of Cold-Cathode Vacuum Gauges." *R&D Magazine,* October 2000, pp. 35–36.

Brucker, Gerardo. "Which Bayard-Alpert Gauge Is Best for You?" *R&D Magazine,* February 2000, pp. 69–72.

Harris, Nigel. *Modern Vacuum Practice,* McGraw-Hill International, New York, NY, 1989.

O'Hanlon, John. *A User's Guide to Vacuum Technology,* 3rd Edition, Wiley Interscience, Hoboken, NJ, 2003.

Product Catalog 2000, Varian Vacuum Technologies, Lexington, MA.

Tompkins, Harland G. *Vacuum Technology: A Beginning,* American Vacuum Society, New York, NY, 2002.

PROBLEMS

1. Determine the effective pumping speed needed for a turbo pump to maintain a pressure of 1×10^{-5} torr in a rectangular chamber made of anodized aluminum and having the following dimensions: 15 cm wide by 10 cm high by 20 cm long. Assume that the outgassing rate for anodized aluminum is 3×10^{-5} torr liters per second per square centimeter.

2. Determine the effective pumping speed required for a process that requires 15 sccms of process gas at a process pressure of 20 millitorr.
3. What is the nitrogen pumping speed for a Varian Turbo-V2000HT pump? What is the rotational speed for this turbo pump? What is the recommended dry backing pump for this turbo pump? Use the Internet to find this information.
4. Determine the net conductance of a straight tube 30 cm long and 4.0 cm in diameter operating in the high-vacuum regime.
5. A cylindrical stainless steel tank is evacuated using a turbomolecular pump backed by a rotary vane pump. The tank is 25 cm diameter and 40 cm in height. Assume an outgassing rate for stainless steel of 5×10^{-8} torr liters/sec cm2 and a pumping speed for the turbo pump of 90 liters/second for nitrogen. Estimate the ultimate pressure that can be achieved by pumping the tank down from atmosphere. Assume that water desorption is not a significant factor.

LABORATORY ACTIVITIES

Activity 1: *Outgassing*

Equipment Needed: A high-vacuum pumping system with a minimal volume chamber (Tee, thermocouple or Pirani rough vacuum gauge, Bayard-Alpert high-vacuum gauge) and stopwatch. Vacuum system should operate automatically—that is, the controller should automatically start the high-vacuum pump. The only manual operation necessary might be turning on the Bayard-Alpert gauge at the appropriate pressure.
Paper towel and a small amount of water. Ruler. Scissors.

Objective: To observe the effect of outgassing on the pump-down curve for a high-vacuum system.

Pre-Lab Activity: None

Laboratory Procedure: The chamber has been vented to the atmosphere and is at atmospheric pressure.

1. Close the vent valve, calibrate the rough vacuum gauge to atmospheric pressure (760 torr).
2. Perform a high-vacuum pump down, taking pressure readings every 15 seconds. Take data until base pressure is reached.
3. Plot the pressure versus time data on semi-log graph paper.
4. Vent the chamber to atmosphere and immediately start another high-vacuum pump down.
5. Take data at the same time intervals as during the first pump down until base pressure is reached.
6. Plot the pressure versus time data on the same semi-log graph paper as used in Step 3. Note the difference between the two pump-down curves.
7. Cut a 2 cm^2 piece of dry paper towel.

8. Vent the chamber, open the chamber, and place the dry paper towel in the chamber. Place the paper towel in a location away from the throat of the high-vacuum pump so it won't fall into the pump. Close the chamber.
9. Perform another high-vacuum pump down, taking pressure readings at the same intervals as the first two pump downs. Continue taking data for the same total time interval as the first two pump downs.
10. Plot the pressure versus time data on the same semi-log graph as the first two pump downs. Compare the third pump-down curve to the first two pump-down curves.
11. Vent the chamber and remove the paper towel. Apply a small amount of water to slightly wet the towel. Do not saturate it. Place the damp paper towel in the chamber and close the chamber.
12. Perform a fourth pump down, taking pressure readings at the same intervals and for the same total time period as the previous pump downs.
13. Plot the pressure versus time data for the fourth pump down on the same semi-log graph as the other pump downs. Compare the fourth pump-down curve to the previous curves.
14. Vent the chamber and remove the paper towel. Close the chamber. Observe the wetness of the paper towel.
15. Summarize your observations and analyze the graph with the four pump-down curves. Write a concise summary explaining why the pump-down curves are different.

Activity 2: *High-Vacuum Pump Down: Manual Operation*

Equipment Needed: A high-vacuum pumping system turbomolecular pump backed with a diaphragm pump, with a bell jar chamber, thermocouple or Pirani rough vacuum gauge, Bayard-Alpert high-vacuum gauge). The rough vacuum and high-vacuum pumps should be manually operated. Stopwatch and semi-log graph paper.

Objective: To obtain a pump-down curve for a high-vacuum system.

Pre-Lab Activity: By making measurements of the dimensions of the chamber, attached piping, and other components exposed to the chamber, estimate the internal surface area of each material used to construct the chamber and associated components. Using the surface areas, approximate outgassing rates for the materials, and pumping speed, estimate the ultimate pressure that can be achieved in the chamber at the end of a high-vacuum pump down.

Laboratory Procedure: The chamber has been vented to the atmosphere and is at atmospheric pressure.

1. Close the vent valve, calibrate the rough vacuum gauge to atmospheric pressure (760 torr). Close the roughing valve, foreline valve, and high-vacuum gate valve.
2. Perform a rough vacuum pump down until the cross-over pressure is reached.
3. Close the roughing valve. Open the foreline valve. Start the high-vacuum pump and let it reach its operating speed.
4. Reset the stopwatch to zero. Open the high-vacuum gate valve and start the stopwatch. Take pressure measurements at 15-second intervals for the first

several minutes. Lengthen the measurement intervals as the pressure decrease slows. Take pressure measurements for 1–2 hours.

5. Let the vacuum system run overnight. Take a pressure reading and record the elapsed time of the pump down.
6. Plot the pressure versus time data on semi-log graph paper. What is the shape of the pump-down curve?
7. Compare the last pressure reading to your predicted ultimate pressure. How close are the two values? How can you account for any difference in the two values?

CHAPTER 6

Partial Pressure Analysis Using Residual Gas Analyzers

6.1 INTRODUCTION

The primary application of partial pressure analysis is to determine the composition of gas in a vacuum system. This is commonly performed by a *mass spectrometer*. The mass spectrometer most often used today for partial pressure analysis is the quadrupole residual gas analyzer, or RGA.

Why would we want to know the partial pressure for specific gases in a vacuum system? One reason is that it enables us to analyze and monitor process chemistries in the process chamber by analyzing the by-products of the reaction. Another is to detect impurities in the chamber and monitor gas fills during a manufacturing process. A third might be to detect leaks using a tracer gas such as helium.

What is the difference between a vacuum pressure gauge such as an ionization gauge and an RGA? A vacuum gauge measures total pressure. Since the residual gas within a vacuum system is almost always a combination of gases, vacuum gauges measure the sum of the partial pressures exerted by each specific gas. For example, if there is room air in the chamber, the vacuum gauge reading will be approximately equal to the sum of the partial pressures of nitrogen, oxygen, and water vapor, the major gaseous components of room air. On the other hand, a *residual gas analyzer* measures the partial pressure of each specific gas being removed from the chamber. From the RGA display, we can tell which gases are present and in what relative amounts.

The RGA has evolved over the years from a basic leak detector to a sophisticated instrument used peripheral to or integrated into complex manufacturing tools. Interfaced to a computer, an RGA can supply the right information to identify and solve such process problems as contamination, process recipe deviations, gas-flow issues, and maintenance issues. Correlating RGA data with process conditions is a vital part of using RGAs in the research and development, implementation, and manufacturing phases of a product's lifetime. It has made the RGA an indispensable tool in both the research lab and the factory.

6.2 CONSTRUCTION AND OPERATION OF A QUADRUPOLE RGA

Today's RGA is made up of three basic components: an ionizer, an ion filter, and an ion detector. Of course, the RGA must also have an electronics control unit that takes the signal from the ion detector and sends it to a computer where the control software processes the signal and presents the information on the monitor screen. As our example of an RGA, we have selected the Stanford Research Systems RGA-200 Residual Gas Analyzer manufactured by Stanford Research Systems, located in Sunnyvale, California. The discussion that follows refers to a quadrupole residual gas analyzer as simply an RGA.

The quadrupole probe and electronics control unit (ECU) of the analyzer are shown in Figure 6.1. The quadrupole probe consists of an ionizer and an ion filter. The ion detector is located at the opposite end of the quadrupole probe from the ionizer and connects to the ECU. The RGA-200 mounts directly to a standard 2.75-inch CF port on the vacuum system.

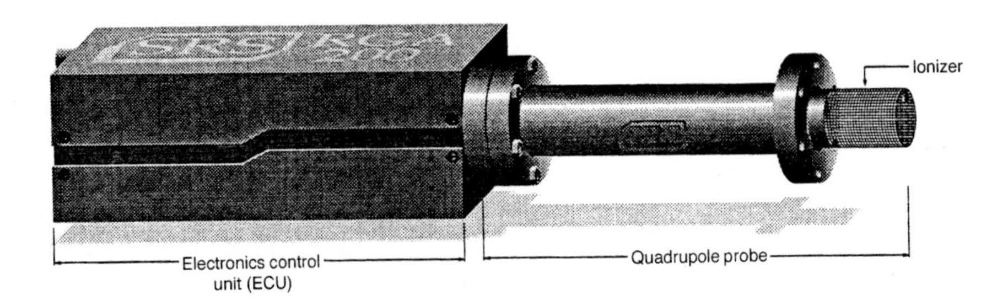

FIGURE 6.1
Components of RGA-200.
Source: SRS Operating Manual, pp. 3–2.

6.2.1 The Ionizer

Let us begin at the ionizer and work our way back to the ECU. The principal parts of the ionizer include the repeller, the anode grid, the filament, and the focus plate (see Figure 6.2). The role of the ionizer, as its name implies, is to ionize some of the gas molecules in the chamber and focus a beam of ionized gas atoms and molecules into the ion filter.

The filament is made from oxidation-resistant thoria-coated iridium wire. Thoria-coated iridium wire can operate at low temperatures and can be exposed to atmosphere without the risk of burning up. The RGA-200 has two cylindrical filaments that operate simultaneously under normal operation so that service is not interrupted if one burns out.

The operation of the ionizer unit is similar to a Bayard-Alpert gauge, except that there is an electron repeller and no central wire collector. The filament is the source of the electrons used to ionize the gas atoms and molecules. When heated to incandescence, it produces thermionically emitted electrons. The emitted electrons are repelled by the negative

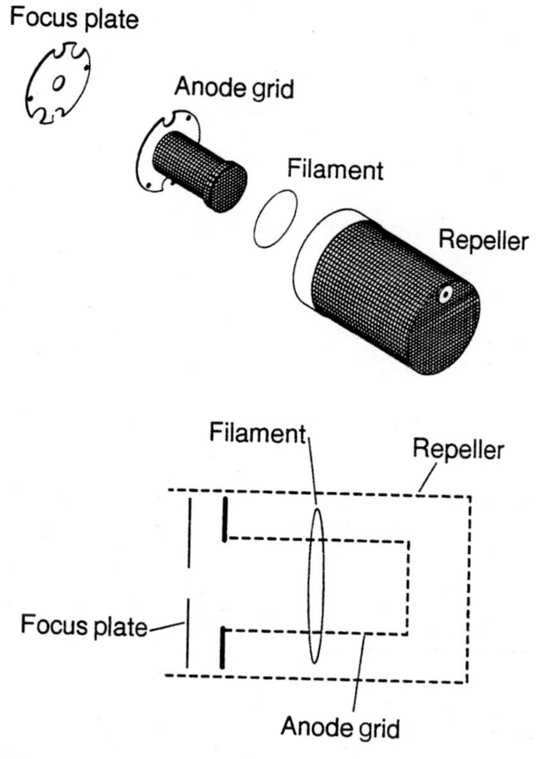

FIGURE 6.2
Components and schematic diagram of ionizer.
Source: SRS Operating Manual, pp. 3–3, 3–4.

potential on the filament and are accelerated toward the positive potential on the anode grid. Because of the open wire mesh design of the grid, most of the electrons do not strike the anode immediately, but pass through and enter a region between the grid and the repeller. It is in this region that electron impact ionization occurs. Those electrons that do not strike the grid or ionize a gas molecule pass into the space between the grid and the repeller. Here they come under the influence of the negative potential on the repeller, slow down, and are reaccelerated back toward the grid, thus getting another chance to ionize a gas molecule. The circulation of electrons between the filament and the repeller increases the ionization efficiency of the electron current relative to single-pass configurations.

Positive ions produced within the anode grid volume are extracted from the ionizer by an electric field produced by the difference in voltage bias between the anode grid and the focus plate. The focus plate is biased negatively with respect to ground.

Electron impact ionization within the RGA requires low pressures for efficient production of electrons and ions. RGAs are best suited for operation in the high- and ultrahigh-vacuum regimes (pressures from 10^{-4} to 10^{-14} torr). However, RGAs can be nonlinear above 10^{-5} torr due mainly to space charge effects, and for this reason they are normally operated at pressures of 10^{-5} torr or below. The ECU will prevent the filament from turning on if the pressure is too high and will turn off the filament if the pressure rises above the maximum operating pressure for the RGA.

The ECU contains all the necessary high voltage and current supplies needed to bias the ionizer's electrodes and establish the selected electron emission current. Four parameters that determine the efficiency of the ionizer are controlled: electron energy, ion energy, electron emission current, and focusing voltage.

The average electron energy, expressed in electron volts, or eV, is equal to the voltage difference between the filament and the anode grid. This voltage difference ranges from 25 eV to 105 eV in the RGA-200. To produce ionization of gas molecules by electron bombardment, the electrons must have a certain minimum kinetic energy, called the *ionization potential.* Ionization potentials differ for each gas. Above the ionization potential, the ionization efficiency increases linearly with the electron energy until a maximum is reached. For most gas molecules, this maximum is in the range of 50 eV to 100 eV. The default setting for the RGA-100 is 70 eV, and it has a range of settings from 25 eV to 100 eV.

The *ion energy* is the kinetic energy of the ions as they move down the ion filter, and it too is expressed in eV. The ion energy is equal to the bias voltage on the anode grid. The ion energy setting affects the magnitude of the ion signals collected by the ion detector (the sensitivity of the RGA) and limits the ultimate resolution of the mass filter. In other words, the anode grid sets the kinetic energy of the ions as they enter the ion filter. The ion energy determines the time spent in the ion filter and thus the resolution that can be obtained. The more energy the ion possesses, the faster it will travel through the ion filter. Conversely, the less energy the ion possesses, the slower it will travel through the ion filter.

The electron emission current is the electron current from the filament to the grid. The RGA-100's emission current has a default setting of 1.00 mA and a range of 0 to 3.5 mA in increments of 0.02 mA. The electron emission current is tightly controlled by a feedback control loop which dynamically adjusts the filament temperature to keep the total emission current constant.

The focus plate's negative potential can be adjusted from 0 V to −150 V and is selected to optimize the ion signals. The focus plate serves the dual purpose of drawing ions away from the anode grid while containing ionizing electrons within the source. The default setting for the RGA-100's focus plate voltage is −90 V.

6.2.2 The Ion Filter

The RGA's ion filter is an electrodynamic quadrupole mass filter constructed out of four electrically conductive cylindrical rods, as shown in Figure 6.3. The cylindrical rods are held in quadrature and parallel to each other.

The operation of the quadrupole mass filter can be visualized in the following way. A postive DC voltage with superimposed sinusoidal RF voltage is connected on one pair of rods (X–Y plane). A negative DC voltage with superimposed sinusoidal RF voltage, 180 degrees out of phase with the RF voltage on the first set of rods, is connected on the other pair of rods (Y–Z plane). The two potentials can be represented by the following expressions:

$$V_x = +V_{DC} + V_m \cos \omega t$$
$$V_y = -V_{DC} - V_m \cos \omega t$$

where V_{DC} is the magnitude of the DC voltage applied to either pair of rods,
V_m is the amplitude of the RF voltage applied to either pair of rods,
and ω is the angular frequency of the RF voltage.

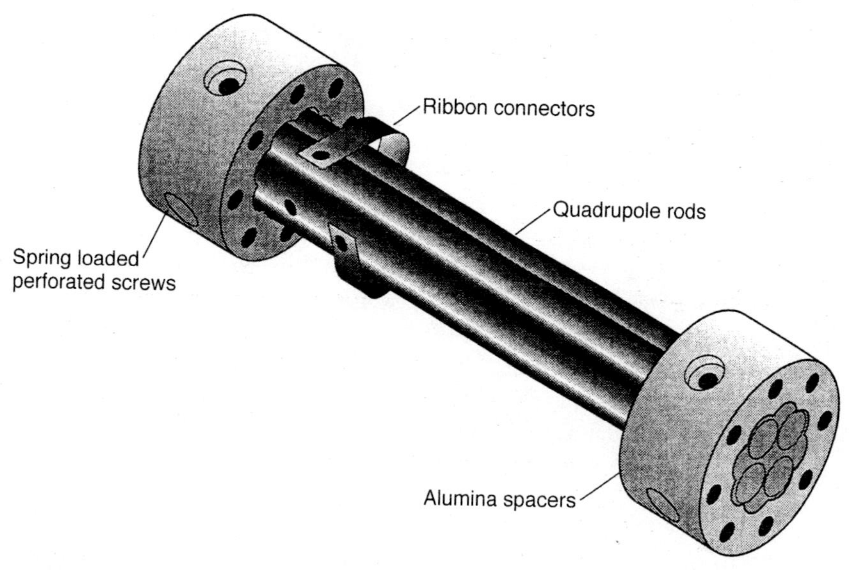

FIGURE 6.3
Ion filter.
Source: SRS Operating Manual, Fig. 5, p. 3–7.

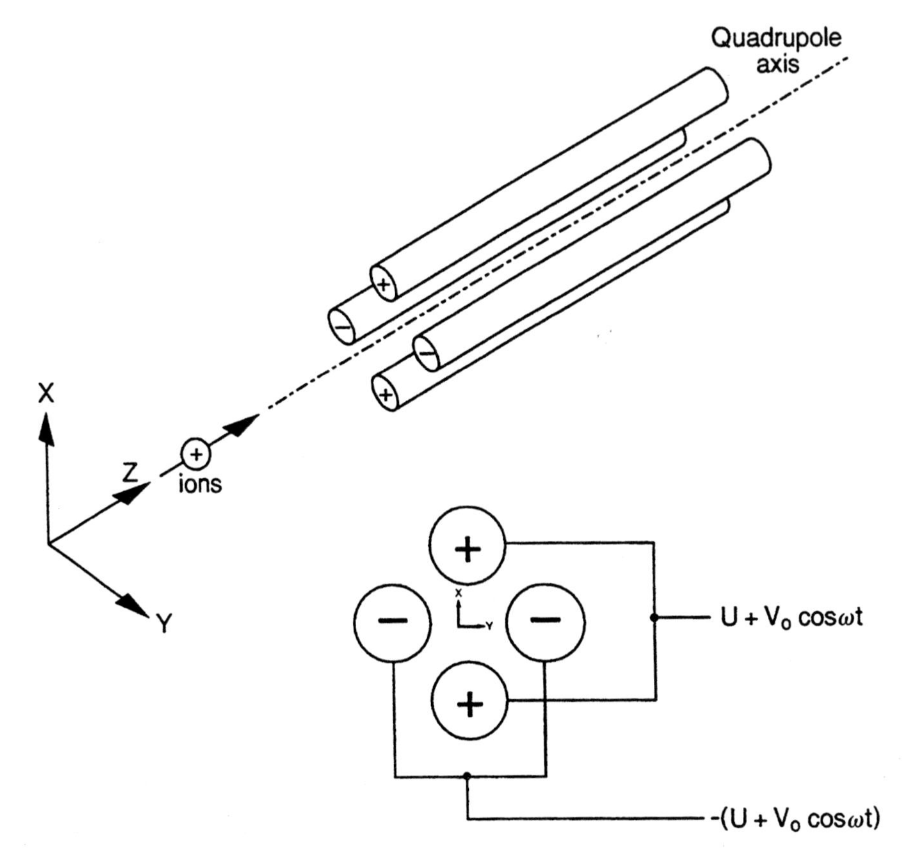

FIGURE 6.4
Voltage applied to rods in quadrupole mass filter.
Source: SRS Operating Manual, Fig. 6, p. 3–8.

During operation, ions enter the mass filter along the Z-direction and begin to oscillate in the X- and Y-directions. The mass-to-charge ratio of the ion determines the trajectory of travel as the ion moves down the mass filter toward the ion detector at the far end. Ions that have an unstable trajectory collide with the rods and are neutralized before they can reach the ion detector. Ions with stable trajectories reach the ion detector and are counted. If the amplitude of the DC and RF voltages are varied simultaneously, an entire mass spectrum can be scanned. (See figure 6.4.)

The filtering process can be explained in the following manner. For the X-direction, light ions stay in phase with the RF signal, gain energy, and oscillate or spiral in increasingly large amplitudes until they encounter one of the rods and are neutralized. Therefore, the X-direction serves as a high-pass mass filter that only transmits high masses to the other end of the mass filter. On the other hand, in the Y-direction, heavy ions will be unstable because of the defocusing effect of the DC component, but lighter ions will follow stable trajectories. Thus, the Y-direction serves as a low-pass mass filter. The overlay of the effect of these two filters produces a bandpass response that allows ions having a single mass-to-charge ratio to reach the ion dectector.

6.2.3 The Ion Detector

The positive ions that successfully traverse the entire length of the quadrupole mass filter are focused toward the ion detector by an exit aperture held at ground potential. The detector measures the ion currents directly using a Faraday Cup, or optional electron multiplier detector, that measures an electron current proportional to the ion current.

As shown in Figure 6.5, the Faraday Cup is a small stainless steel metal bucket located on-axis at the end of the quadrupole mass filter. The exit plate shields the Faraday Cup from the intense DC and RF fields in the mass filter. Positive ions enter the grounded detector, strike the metal wall, and are neutralized by electron transfer from the metal to the ion. The electrons given up in this process produce a current that has the same magnitude as the incoming ion current. The magnitude of the current is very small and on the order of 10^{-9} to 10^{-15} amperes for pressures in the range of 10^{-5} to 10^{-11} torr. The advantages of the Faraday Cup include simplicity of design, stability, large dynamic range, and lack of mass discrimination. All ions are detected with the same efficiency regardless of their mass.

6.2.4 Operating Modes

Several modes can be used to display data collected by the RGA, among them the analog mode display, histogram mode display, table mode display, pressure versus time mode display, leak test mode display, annunciator mode display, and library mode display. The foregoing modes are depicted in Figure 6.6. They may not all be available on RGAs from other manufacturers.

The *analog mode* is a spectrum analysis mode common to all RGA's. The mass-to-charge ratio is plotted on the X-axis, and the ion current amplitudes corresponding to each mass-to-charge ratio are plotted on the Y-axis.

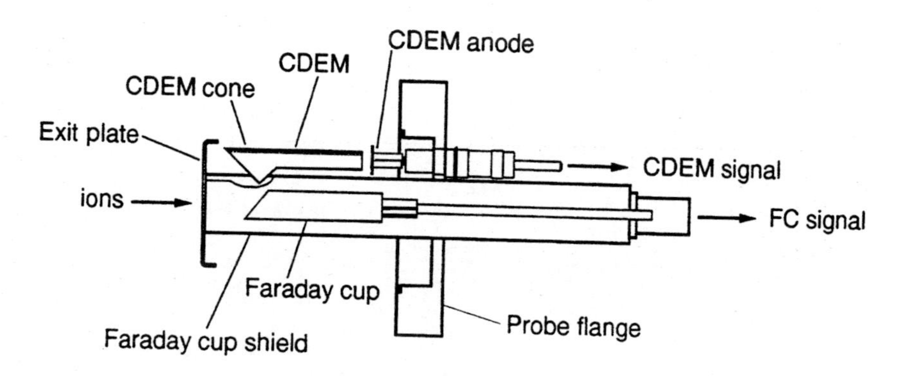

FIGURE 6.5

Faraday Cup ion detector.

Source: SRS Operating Manual, Fig. 8, p. 3–13.

(a) Analog Mode

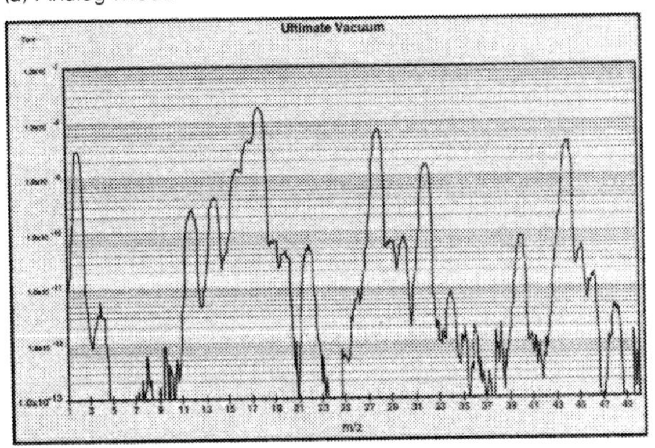

(b) Histogram Mode

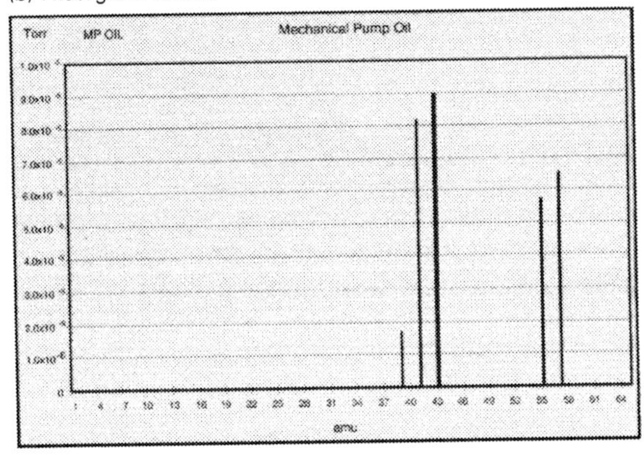

(c) Table Mode

RGA Table Scan

Ch#	Name	Mass	Value	Alarm	Speed	Cal	CEM
1	Hydrogen	2	3.8E-07	NORMAL	1	1.00	OFF
2	Water	18	7.1E-08	HIGH	1	1.00	OFF
3	Nitrogen	28	1.4E-08	HIGH	1	1.00	OFF
4	Oxygen	32	4.5E-10	NORMAL	3	1.00	ON
5	CO2	44	3.4E-11	NORMAL	3	1.00	ON
6	Oil	55	1.6E-12	NORMAL	3	1.00	ON
10	Floor	21	1.1E-13	NORMAL	1	1.00	ON

FIGURE 6.6
RGA display modes.
Source: SRS Product Catalog.

(d) Pressure versus Time Mode

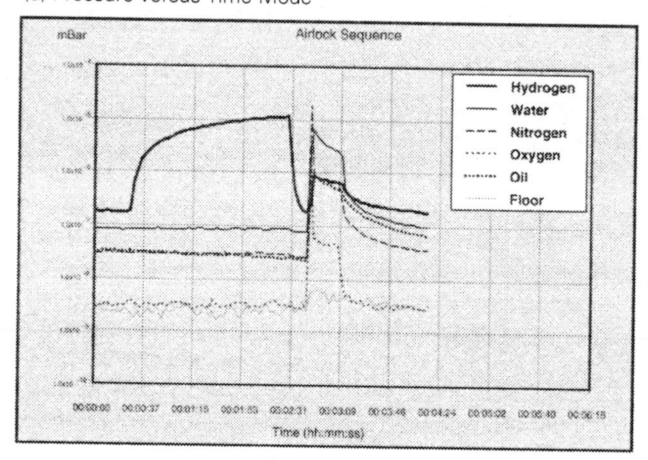

(e) Leak Test Mode

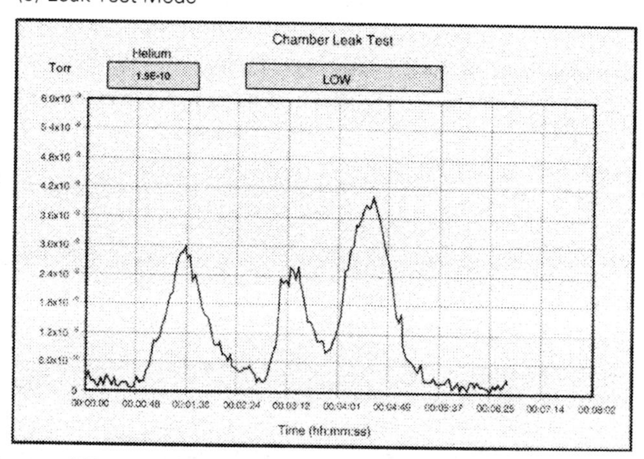

(f) Annunciator Mode

File Edit View Mode Scan Graph Head Utilities Window Help

Hydrogen	Water	Nitrogen	Oxygen	Oil
1.1E-07	5.6E-08	1.4E-08	2.9E-09	1.4E-08
NORMAL	HIGH	NORMAL	NORMAL	EXTREME HIGH

Floor				
3.0E-09	OFF	OFF	OFF	OFF
NORMAL				

FIGURE 6.6 *(Continued)*

The *histogram mode* display is similar to the analog mode display. The histogram display is essentially a bar graph version of the analog mode display. That is, the peak is plotted as a bar at the appropriate mass-to-charge ratio.

The *table mode* display presents a tabular readout of preselected gases along with alarm-level warnings. The gas masses, names, and parameters are user selected. In this mode, the partial pressure value is acquired directly from the RGA head by individually querying the partial pressure for the appropriate mass.

The *pressure versus time mode* is a scroll graph of user-selected gas mass-to-charge ratios on the same plot. The graph scrolls to the left as the data fill the screen, and the old data are saved in a history buffer for review at any time.

The *leak test mode* provides the most effective way to study the behavior of a single gas. The display is a scroll graph that monitors the tracer gas trend over a period of time. Instantaneous partial pressure readout, bar meter, and alarm messages may also be available. In the leak test mode, the partial pressure for the specific tracer gas is acquired by querying the RGA head for only that specific mass-to-charge ratio.

The *annunciator mode* provides an effective way to visually monitor gas-warning levels from a distance. The display is composed of large green panels representing the gas and alarm levels. When an alarm level is reached, the panel color changes to bright red and the appropriate alarm message is shown.

The final mode is the *library mode.* This mode displays the selected library gas fragmentation pattern as a histogram graph. Using the split-screen mode of operation, analog mode displays can easily be compared to fragmentation patterns stored in the library to identify unknown gases.

6.3 PARTIAL PRESSURE MEASUREMENT

When an electron with sufficient energy to cause ionization collides with a gas molecule, several different ions or fragments are created. The mass-to-charge ratios of these ions are unique for each gas species, and the pattern of fragments is called a *cracking pattern* for the gas. Using an RGA, the known cracking patterns for various gases can be used to determine the composition of the gas in the process chamber.

The various fragments in the cracking pattern are created by simple ionization, dissociative ionization, multiple ionization, and isotopic mass differences. To show how this fragmentation process works, consider what happens when a diatomic nitrogen gas molecule collides with an electron with sufficient energy to cause ionization.

The most common reaction is simple ionization of the diatomic nitrogen gas molecule:

$$N_2 + e^- \rightarrow N_2^+ + 2e^-$$

The ion created is N_2^+ having a mass of 28 atomic mass units (amu) and a charge of +1 or just plain 1. The mass-to-charge ratio for this fragment is 28/1, or 28.

The cracking pattern for nitrogen also shows a fragment having a mass-to-charge ratio of 14. This is a singly charged nitrogen atom. The reaction can be written as

$$N_2 + e^- \rightarrow N + N_2^+ + 2e^-$$

The ion N^+ has a mass of 14 amu and a charge of +1. Thus, the mass-to-charge ratio is 14/1, or 14. This fragment is not as common as N_2^+. For every 100 N_2^+ ions, only seven N^+ ions are created.

The third most common fragment given for nitrogen gas has a mass-to-charge ratio of 29. This occurs when one of the two nitrogen atoms in the diatomic gas molecule is an isotope of nitrogen having a mass of 15 amu. If the diatomic nitrogen gas molecule containing an isotope is ionized, the resulting ion will have a mass of 29 amu and a charge of +1, a mass-to-charge ratio of 29.

For nitrogen gas, besides mass-to-charge ratios of 28, 14, and 29, there are three other fragments. These three fragments have relative abundance values much less than 1. One of these fragments has a mass-to-charge ratio of 7, corresponding to N^{++}. Another fragment has a mass-to-charge ratio of 15, corresponding to $^{15}N^+$, a singly charged isotope of nitrogen. And finally, a fragment can have a mass-to-charge ratio of 30, corresponding to $^{15}N_2^+$, a singly charged diatomic molecule consisting of two nitrogen isotopes.

A shorthand way of representing the cracking pattern for a substance is to list the mass-to-charge/relative abundance ratios for the various fragments, starting with the most abundant and ending with the least abundant. Only fragments with a relative abundance greater than 1 are listed. For nitrogen, the cracking pattern can be represented as

Nitrogen	28/100	14/7	29/1

where the numerator is the mass-to-charge ratio, and the denominator is the relative abundance. Note that the most common fragment is always given a relative abundance of 100. This same information can be presented in graphical form. Figure 6.7 shows a histogram plot for nitrogen gas.

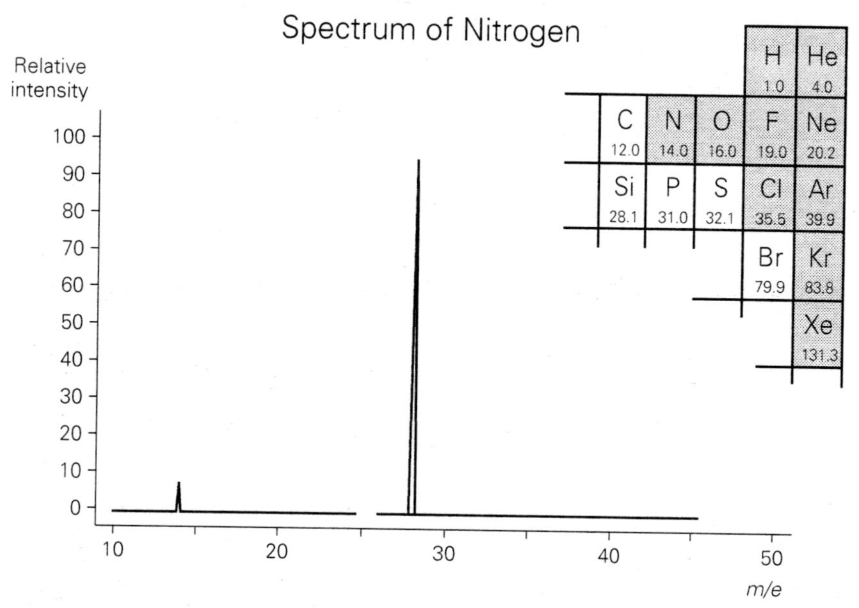

FIGURE 6.7
Histogram plot for nitrogen gas.

EXAMPLE 6.1 Identify the fragments in the cracking pattern for water.

Mass-to-charge ratio	Relative abundance
18	100
17	25
1	6
16	2
2	2

Solution

The most common fragment for water, H_2O, is a singly charged water molecule, H_2O^+, with a mass-to-charge ratio of 18. The next most prevalent fragment with a mass-to-charge ratio of 17 is HO^+. The fragment with a mass-to-charge ratio of 16 is a singly charged oxygen atom, O^+. The rarely occurring fragments for water are the singly charged hydrogen atom, H^+, and a singly charged hydrogen molecule, H_2^+. The three most abundant fragments are shown in Figure 6.8.

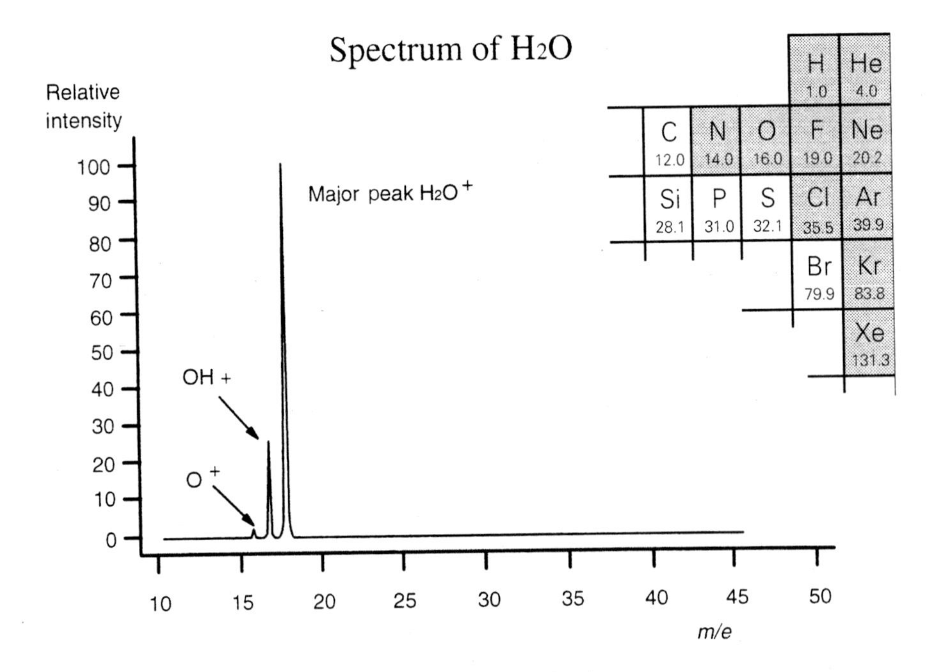

FIGURE 6.8
Histrogram plot for H_2O.
Source: MKS—"Partial Pressure Techniques & Residual Gas Analyzer, p. 44.

6.4 INTERPRETING MASS SPECTRA

When interpreting the mass spectra from an RGA, it is very important to correctly identify the mass-to-charge ratios of each peak in the mass spectrum. The mass-to-charge ratios are plotted along the horizontal axis of the graph produced by the RGA software. Then the residual gases producing the spectrum can be identified.

In general, we can divide gases into two categories, inorganic gases and organic gases. *Organic gases* are composed of or contain carbon and hydrogen atoms—for instance, CH_4 (methane) and C_2H_6 (ethane). All other gas molecules can be classified as *inorganic molecules*—for instance N_2 (nitrogen), O_2 (oxygen), and CO_2 (carbon dioxide).

6.4.1 Inorganic Gases

Let's first consider inorganic molecules. If a vacuum system has been vented to the atmosphere, it contains air which is composed of inorganic gases, mainly N_2, 0_2, Ar, and H_20 (see Figure 6.9). The residual constituents in most vacuum systems will also be inorganic gases such as H_2 and CO. All of these gas molecules are simple molecules consisting of one, two, or three atoms. In consequence, their fragmentation patterns will be rather simple, and the gases should be easy to identify in the mass spectrum. There is a strong tendency for simple

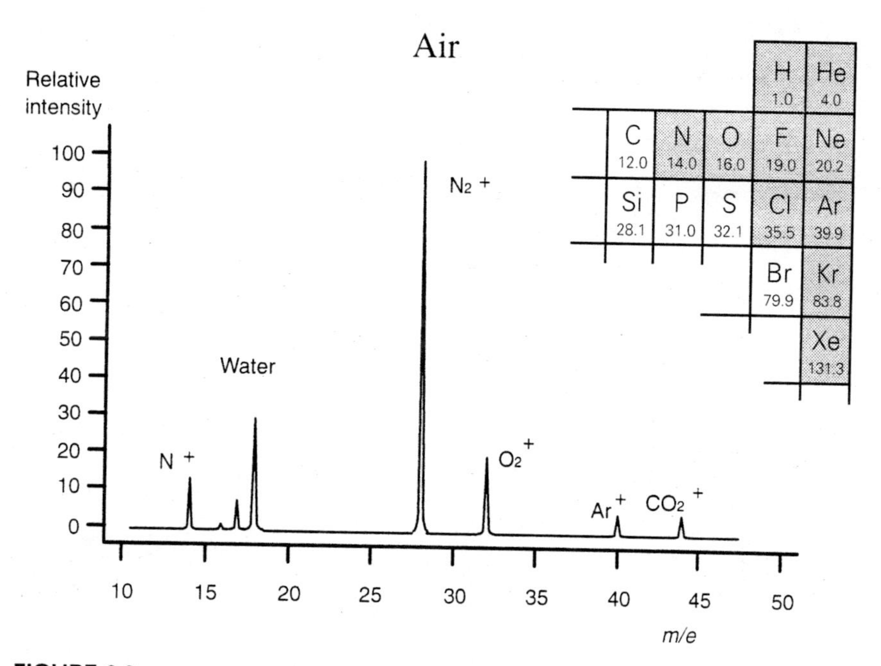

FIGURE 6.9
Spectra for air.
Source: MKS Instruments, "Partial Pressure Techniques and the Residual Gas Analyzer," p. 45.

inorganic gas molecules to have even mass numbers, and that is the situation in this case. The most common fragment is produced by simple ionization of the parent molecule, for example N_2^+, O_2^+, Ar^+, and H_2O^+. So, as a general rule, the parent peak of most inorganic gases will be the largest. It will generally be found at an even mass-to-charge ratio.

Analysis of the spectra of inorganic gases indicates that the more massive fragments will tend to carry the unbalanced charge. For example, in the case of water, consider the following two reactions:

$$H_2O + e^- \rightarrow OH^+ + H + 2e^-$$
$$H_2O + e^- \rightarrow OH + H^+ + 2e^-$$

The relative abundance of OH^+ is 27.01. Compared to the relative abundance of H^+ of 0.1, this indicates that the upper reaction is much more likely to occur, since OH can more easily distribute the charge than the less massive hydrogen atom.

The observation that more massive fragments tend to carry the unbalanced charge is less obvious in the case of carbon dioxide. Consider the two reactions

$$CO_2 + e^- \rightarrow CO^+ + O + 2e^-$$
$$CO_2 + e^- \rightarrow CO + O^+ + 2e^-$$

Table 6.1 lists the spectra for selected inorganic gases. For those gases with more than six fragments, only the six most abundant fragments are listed.

TABLE 6.1
Spectra for inorganic gases.

Substance	1	2	3	4	5	6
Air (N_2, O_2, Ar, etc.)	28/100	32/27	14/6	16/3	40/1	
Ammonia (NH_3)	17/100	16/80	15/8	14/2		
Argon (Ar)	40/100	20/10				
Carbon dioxide (CO_2)	44/100	28/11	16/9	12/6	45/1	22/1
Carbon monoxide (CO)	28/100	12/5	16/2	29/1		
Carbon tetrachloride (CCl_4)	117/100	119/91	47/51	82/42	35/39	121/29
Carbon tetrafluoride (CF_4)	69/100	50/12	31/5	19/4		
Helium (He)	4/100					
Hydrogen (H_2)	2/100	1/5				
Hydrogen sulfide (H_2S)	34/100	32/44	33/42	36/4	35/2	

Krypton (Kr)	84/100	86/31	83/20	82/20	80/4	
Neon (Ne)	20/100	22/10	10/1			
Nitrogen (N_2)	28/100	14/7	29/1			
Oxygen (O_2)	32/100	16/11				
Silane (SiH_4)	30/100	31/80	29/31	28/28	32/8	33/2
Silicon tetrafluoride (SiF_4)	85/100	87/12	28/12	33/10	86/5	47/5
Water (H_2O)	18/100	17/25	1/6	16/2	2/2	
Xenon (Xe)	132/100	129/98	131/79	134/39	136/33	130/15

• Legend: Mass/Relative Abundance
• The six most intense peaks in the spectrum of each substance.
Source: Inficon Spectra Library Card, 2002.

In the CO_2 spectra, the fragment CO^+ has a relative abundance of 15, while O^+ has a relative abundance of 16. In this case, the difference in masses is not as great, approximately 2:1, as compared to the case of water, where the difference of masses between OH^+ and H^+ is 17:1.

6.4.2 Organic Gases

The main sources of residual organic gases in vacuum systems are solvents, pump oils, contamination such as fingerprints, and materials like plastics. Solvents include alcohols, hydrocarbons, freons, and acetone.

Organic gases are relatively large molecules consisting of carbon and hydrogen. The large number of atoms per molecule usually produce fragmentation patterns that are much more complex than the fragmentation patterns for inorganic molecules. Here are some general rules that are helpful in dealing with organic gases:

- Fragments with odd mass numbers are generally more abundant.
- Spectra have a series of peaks separated in mass by 14 or 15 amu due to the presence of CH_2^+ or CH_3^+ fragments.
- Mass peaks at (57,55,53), (43,41,39), (29,27), or combinations of these are an indication of organic species in the vacuum system.
- Parent peaks for large organic molecules may not appear because they are much heavier than the lighter fragments and may not be detected.

The simplest hydrocarbon is methane, CH_4. The cracking or fragmentation pattern for methane contains seven fragments with a relative abundance of 1 or greater: 16/100 (CH_4^+), 15/85 (CH_3^+), 14/16 (CH_2^+), 13/8 (CH^+), 1/4 (H^+), 12/2 (C^+), and 17/1 (H^+).

A more complex hydrocarbon is isopropyl alcohol, or 2-Propanol, $(CH_3)_2CHOH$. The cracking pattern for isopropyl alcohol (IPA) is: 45/100, 43/16, 27/16, 29/10,

TABLE 6.2
Spectra for organic gases.

Substance	1	2	3	4	5	6
Acetone (CH_3COCH_3)	43/100	15/42	58/20	14/10	27/9	42/8
Chlorobenzene (C_6H_5Cl)	77/100	112/89	51/57	50/48	38/31	114/25
DC 705 pump oil	78/100	76/83	39/73	43/59	91/32	
Fomblin pump oil	69/100	20/28	16/16	31/9	97/8	47/8
PPE pump oil	50/100	77/89	63/29	62/27	64/21	38/7
Ethyl alcohol (CH_3CH_2OH)	31/100	45/34	27/24	29/23	46/17	26/8
Heptane (C_7H_{16})	43/100	41/62	29/49	27/40	57/34	71/28
Hexane (C_6H_{14})	41/100	43/92	57/85	29/84	27/65	56/50
Isopropyl alcohol ($CH_3CH_2OHCH_3$)	45/100	43/16	27/16	29/10	41/7	39/6
Methyl alcohol (CH_3OH)	31/100	29/74	32/67	15/50	28/16	2/2
Mechanical pump oil	43/100	41/91	57/73	55/64	71/20	39/19
Toluene ($C_6H_5CH_3$)	91/100	92/62	39/12	65/6	$45\frac{1}{2}$/4	51/4
Trichloroethane (CH_3CCl_3)	97/100	61/87	99/61	26/43	27/31	63/27
Trichloroethylene (CCl_2CHCl)	95/100	130/90	132/85	97/64	60/65	35/40
Trifluoromethane (CHF_3)	69/100	51/91	31/49	50/42	12/4	
Turbo pump oil	43/100	57/88	41/76	55/73	71/52	69/49

- Legend: Mass/Relative Abundance
- The six most intense peaks in the spectrum of each substance

Source: Inficon Spectra Library Card, 2002.

41/7,19/7, 39/6, 31/6, 42/4, and 59/3. Note the presence of the (43, 41, 39) and (29, 27) combinations of mass peaks in the cracking pattern.

Table 6.2 lists the cracking pattern or spectra for selected organic gases. The spectra for organic gases is more complex than for inorganic gases. Only the six most abundant fragments are listed.

6.5 USES OF A RESIDUAL GAS ANALYZER

Since residual gas analyzers can analyze the gas composition of vacuum systems, they are well suited for a number of applications. One important application is to monitor manufacturing processes. For example, they can obtain time sequences of gas partial pressures during gas fills and detect the endpoints to manufacturing processes. Other applications focus on diagnostic applications, such as detecting impurities and finding leaks. The following examples illustrate some of the uses of an RGA in a manufacturing setting.

Figure 6.10 shows an RGA scan for a manufacturing process where argon is admitted into the chamber as a process gas and ionized to form a plasma. The scan is obtained using the pressure-versus-time mode of operation. The rising edge of the argon curve corresponds to the time when the argon gas supply is turned on, and the falling edge to the time when the gas supply is turned off. Mass-to-charge ratios in the fragmentation

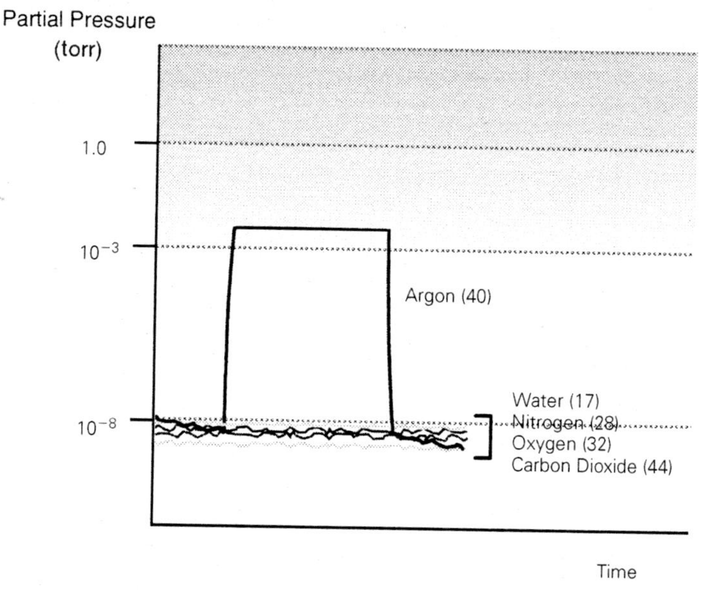

FIGURE 6.10
Normal RGA scan showing argon fill during manufacturing process.
Source: MKS Instruments; "Introduction to Vacuum Gauging Techniques," p. 74.

patterns for water, nitrogen, oxygen, and carbon dioxide are also monitored. In Figure 6.10, the partial pressures for these four gases show no increase during the process cycle.

Now let's consider the same manufacturing process that generated the RGA scan shown in Figure 6.10. However, this time the RGA scan, shown in Figure 6.11, shows an increase in the water partial pressure that corresponds to the interval when the argon is introduced and ionized to form a plasma. In this example, the rise in the partial pressure of water could be due to moisture carried by the wafer into the chamber, and plasma-induced degassing of the water on the wafer causes the observed increase in the partial pressure of water.

Another possible situation is illustrated in Figure 6.12. In this example, the partial pressure of nitrogen increases sharply when the argon gas supply is turned on and falls back to its baseline value when the argon gas supply is turned off. Since the shape of the argon and nitrogen curves are the same, one can conclude that the two events are tied together. One possible cause is that the argon supply is contaminated with nitrogen.

Contamination is a problem in vacuum systems. Consider the RGA scan shown in Figure 6.13. The dominant gas in the chamber is water, as indicated by the peaks at 18 and 17. However, the presence of peaks at 39, 41, 43, 55, and 57 indicates organic

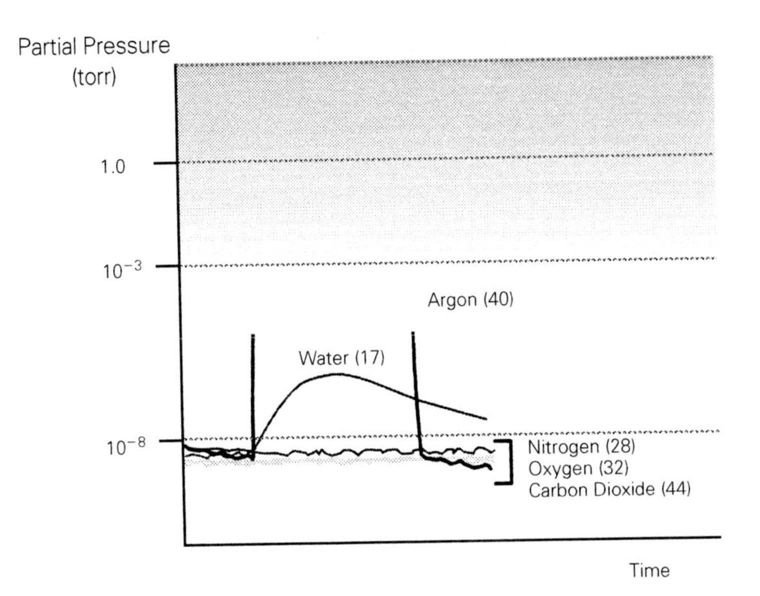

FIGURE 6.11
RGA scan illustrating water contamination.
Source: MKS Instruments; "Introduction to Vacuum Gauging Techniques," p. 75.

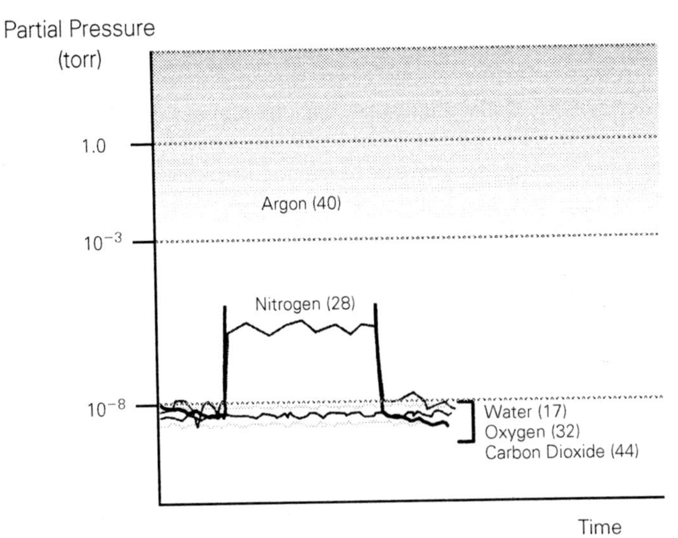

FIGURE 6.12
RGA scan illustrating contamination in a gas supply line.
Source: MKS Instruments; "Introduction to Vacuum Gauging Techniques," p. 76.

(a) RGA scan

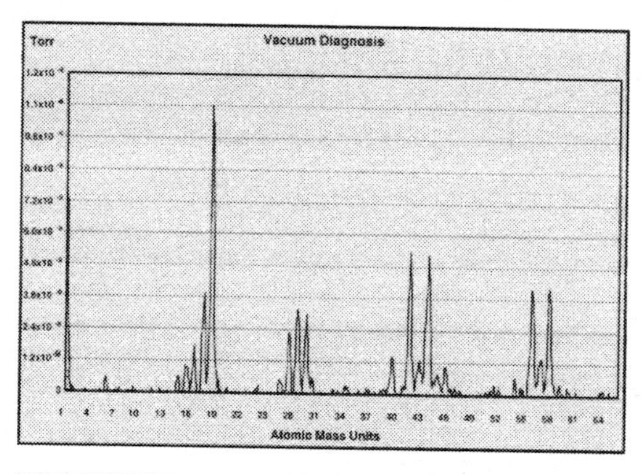

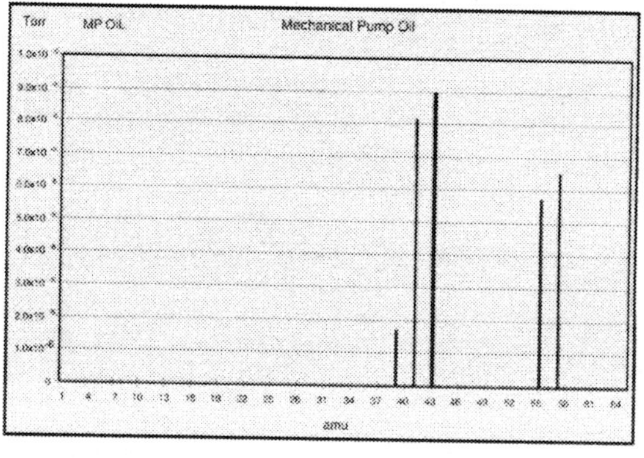

(b) Library scan for mechanical pump oil

FIGURE 6.13
RGA scan indicating mechanical oil contamination.
Source: Stanford Research Systems, Application Note 7.

contaminants in the chamber. These peaks may be caused by mechanical pump oil backstreamed into the vacuum chamber during the process cycle, possibly during the load lock sequence. Two possible causes are improper valve sequencing and a saturated oil trap.

Solvent contamination can also be detected using an RGA. For example, acetone or trichoroethane (TCE) used to clean an oil contaminant can persist in the vacuum system long after cleaning. The RGA scan shown in Figure 6.14 was taken one week after initial contact with the solvent. The major peaks at 97 and 99 and the minor peaks at 61 and 63 are part of the fragmentation pattern for TCE. In this case, the TCE may have permeated into the O-rings in the system during the cleaning process and then slowly outgassed over weeks. To solve the problem, the O-rings should be removed and either baked in an oven to remove the TCE or replaced with new O-rings.

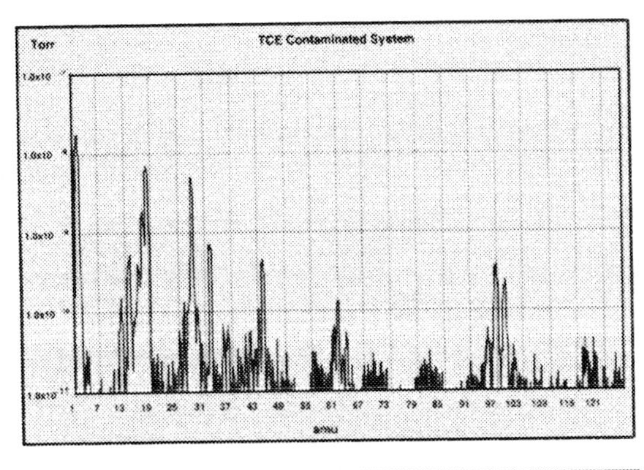

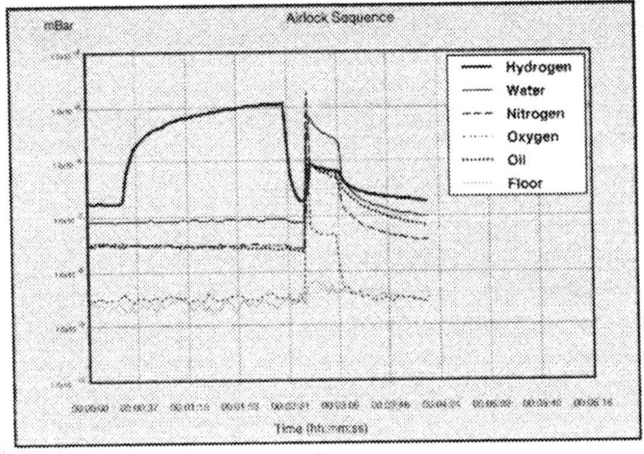

FIGURE 6.14
RGA scan indicating solvent contamination in vacuum system.
Source: Stanford Research Systems, Application Note 7.

SUMMARY

The residual gas analyzer adds a whole new dimension to vacuum system analysis. While vacuum gauges measure the aggregate pressure, RGAs are able to distinguish the partial-pressure fingerprints of individual gases.

The analog and histogram modes plot the partial pressure for each mass-to-charge ratio. Using this graph along with the spectra for gases, users can decode the set of partial pressures into individual gases. The pressure-versus-time mode provides a time history of the partial pressures and can be used to detect changes in the process occurring in the system.

The RGA also provides additional diagnostic capability. In the annunciator mode, the RGA serves as a gas-limit detector. When the alarm level is reached, the panel color for the gas changes to bright red and the appropriate alarm message is shown. In the leak detection mode, the RGA can be set to monitor a tracer gas such as helium, and the output can be an instantaneous partial-pressure readout, bar meter, or alarm message.

BIBLIOGRAPHY

Models RGA100, RGA200, and RGA300 Residual Gas Analyzer: Operating Manual and Programming Reference, Stanford Research Systems, Revision 1.2, Sunnyvale, CA, October 1996.

O'Hanlon, John. *A User's Guide to Vacuum Technology,* 3rd Edition, Wiley Interscience, New York, NY, 2003.

"Partial Pressure Techniques and the Residual Gas Analyzer," MKS Instruments, Wilmington, MA. 1995–98.

"Vacuum Diagnosis with a Residual Gas Analyzer," Application Note #7, Stanford Research Systems, Sunnyvale, CA.

"Will RGAs Replace Ion Gauges?" *R&D Magazine,* http://www.rdmag.com/scripts/, June 2003.

PROBLEMS

1. Compare and contrast the operation of ion gauges and residual gas analyzers.
2. Will RGAs replace ion gauges?
3. For the fragmentation pattern of methyl alcohol, identify the fragments that produce the six most intense peaks in the cracking pattern.
4. For the fragmentation pattern of hydrogen sulfide, identify the fragments that produce the five most intense peaks in the cracking pattern.
5. In the manufacturing process that generated the RGA scan shown in Figure 6.10, a mass-to-charge ratio of 17 was used as an indicator of water rather than the major peak at 18. Why was this done to monitor water in this case?
6. In the manufacturing process that generated the RGA scan shown in Figure 6.12, an increase in nitrogen was observed. Could this result have been produced by an air leak in the nitrogen gas line? Provide supporting evidence for your answer.
7. An unknown gas is analyzed using a residual gas analyzer. The scan is shown in Figure 6.15. Identify the unknown gas and identify the ion responsible for each peak in the scan.
8. The two scans shown in Figure 6.16 were taken two days apart on the same process system after the system had reached base pressure. Figure 6.16(a) was taken when the system was operating normally. The scan in Figure 6.16(b) was taken two days later. What do the two scans tell you about the process system at the time the second scan was taken?

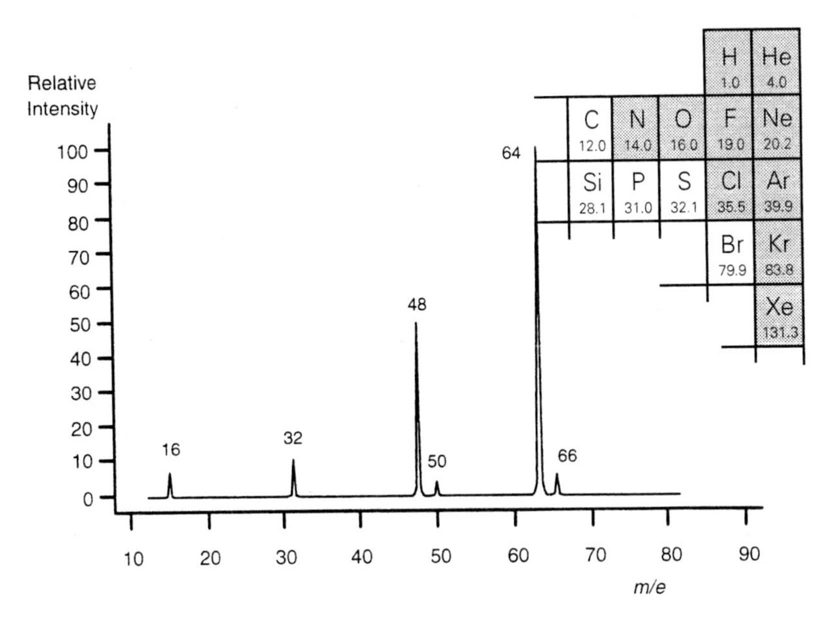

FIGURE 6.15
What is it?

LABORATORY ACTIVITIES

Activity 1: Desorption of Water Study Using an RGA

Objective: To study the rate of desorption of water from the interior surface of a vacuum chamber using a residual gas analyzer or RGA.

Equipment needed: High vacuum system and RGA.

Laboratory procedure: It is assumed that the vacuum system has been vented to the atmosphere and the chamber pressure is at atmospheric pressure. This will insure that the interior surface of the vacuum system will be coated with water molecules.

1. Pump down the vacuum system into the high-vacuum regime and below the maximum operating pressure of the RGA.
2. Configure the RGA to monitor water, nitrogen, and oxygen using the pressure-versus-time mode of operation.
3. When the system pressure drops below the maximum operating pressure of the RGA, turn on the RGA. Begin monitoring the partial pressures of the three main gases: nitrogen, oxygen, and water vapor.
4. Monitor the partial pressures over a period of one hour or longer.
5. From your observations and data, how does the rate of change in partial pressure differ for the three gases?

Activity 2: Leak Detection Using an RGA

Objective: To use a RGA to perform leak checking of a vacuum system.

Equipment needed: High-vacuum system and RGA.

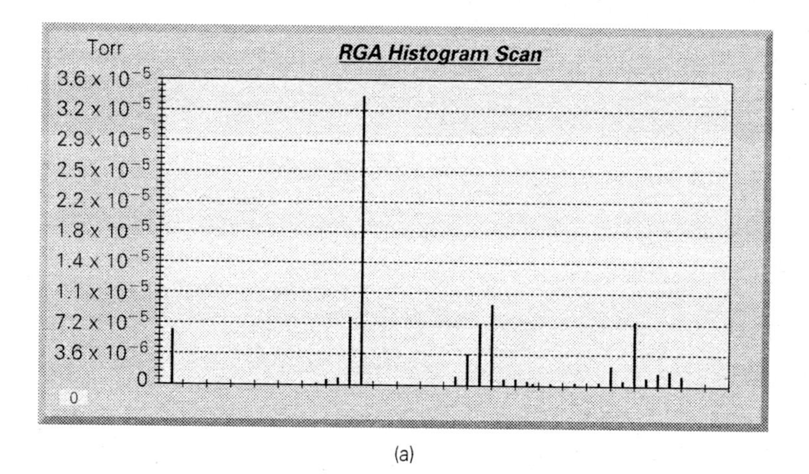

(a)

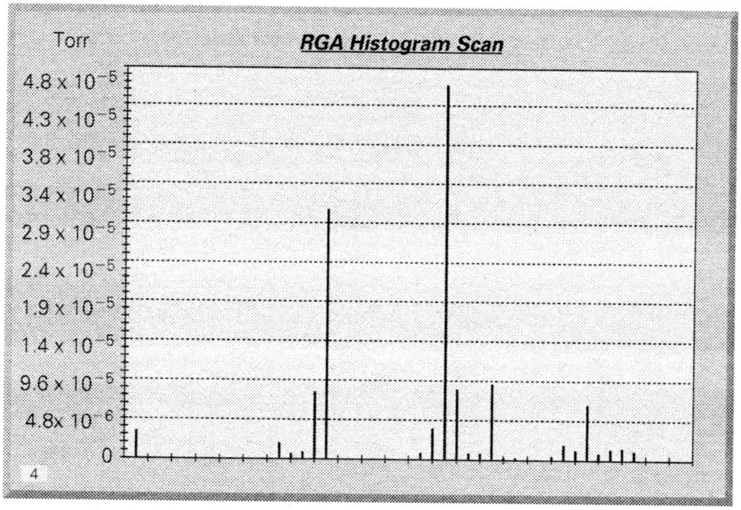

(b)

FIGURE 6.16
RGA scans for the same system taken at different times.

Laboratory procedure:

1. Pump down the vacuum system into the high-vacuum regime and below the maximum operating pressure of the RGA.
2. Configure the RGA for its leak detection mode of operation.
3. Using a gas inlet valve or defective part, create a small leak. Monitor the system pressure with a high-vacuum gauge to insure that the system pressure does not exceed the maximum operating pressure of the RGA.
4. Spray a small amount of helium near the leak and observe the display on the RGA. If a leak is present, the RGA should indicate an increase in the partial pressure of helium.

CHAPTER 7

Leak Detection in Vacuum Systems

7.1 INTRODUCTION

Most of us do not like leaks and wish they would never occur. Unfortunately, that is not the case. They occur all too frequently.

Growing up on a farm, I was constantly dealing with leaks, usually in my bicycle tires. Thorns, rocks, nails, and other objects seemed to have an affinity for my tires. To fix a tire, I had to first remove the tire and the inner tube and then find the leak by submerging the partially inflated inner tube in a tub of water. The bubbles formed by air leaving the hole in the inner tube showed me where I needed to position the tire patch to seal the hole in the inner tube.

Now I own a house, and leaks are not only irritating, but also potentially damaging to my house. Plumbing is a job I do not look forward to. Why can't I buy plumbing components that mate perfectly and never fail? Thank goodness for Teflon plumbing tape.

Leaks occur when a gas or liquid moves across the boundary from the higher-pressure side to the lower-pressure side. In the case of the bicycle tire, the higher-pressure air in the tube escapes through the hole in the tube to the atmosphere. In the case of household plumbing, the higher-pressure water in the pipe escapes through a space in the coupling into the room or through a hole that has rusted through the pipe. In other instances, all it takes is gravity to provide the force to move the water through a leak.

In a vacuum system, the inside pressure is lower than the surrounding atmosphere, so the flow of gas, such as atmospheric air, is from outside the system to inside the system. When there is a water leak in your household plumbing, you can often see the water leaking out of the pipe. In the case of a vacuum system, you cannot see the air entering the system with the naked eye, but using an RGA, it is possible to measure a rise in the partial pressure of the gases making up atmospheric air. We will return to this idea later in the chapter.

Leaks can be classified as either real leaks or virtual leaks. *Real leaks* are leaks that provide a low-conductance pathway for gas to flow from outside the vacuum system to inside the vacuum system. Real leaks may be caused by:

- Porosity through the vacuum chamber wall material
- Cracks in areas of permanent joining, such as welds
- Defective seals due to scratches or contamination on gaskets or defects in the sealing gaskets
- Cracks in feedthrough insulators

Real leaks can be minimized through proper vacuum engineering, fabrication, assembly, and preventative maintenance.

Virtual leaks, in contrast, add to the gas load by allowing an internal volume of trapped gas within a vacuum system to escape into the main volume through a small conductance. Virtual leaks may be due to:

- Trapped volumes such as unvented bolts in blind-tapped bolt holes
- Pores in weld joints and construction materials
- Porosity in coating buildup on vacuum surfaces during repeated processing

Leaks are measured by the rate at which a gas or liquid flows into or out of the leak under certain pressure and temperature conditions. They can be defined either in terms of application, for example 1 ounce of refrigerant R-12 in two years at 70 psi, or in terms of the leak detection method used, such as 2 psi pressure decay in 10 minutes at 40 psi, using the pressure decay method. These definitions are legitimate descriptions of leak rate, but in the United States the generally accepted unit of leak rate for leak detection is std cc/sec; in Europe, the preferred unit is mbar liters/sec. These are used because they specify the actual units of flow rate—that is, mass per unit time. The term *std cc/sec* is an abbreviation for "cubic centimeters of gas per second at standard temperature, 0°C, and standard pressure, 1 atmosphere or 760 torr."

In order to put leak rates into units associated with gas load such as torr-liters/second, the following conversions can be used:

$$1\frac{\text{std cc}}{\text{sec}} = 0.76\frac{\text{torr} \cdot \text{liters}}{\text{sec}}$$

or

$$1\frac{\text{torr} \cdot \text{liter}}{\text{sec}} = 1.3\frac{\text{std cc}}{\text{sec}}.$$

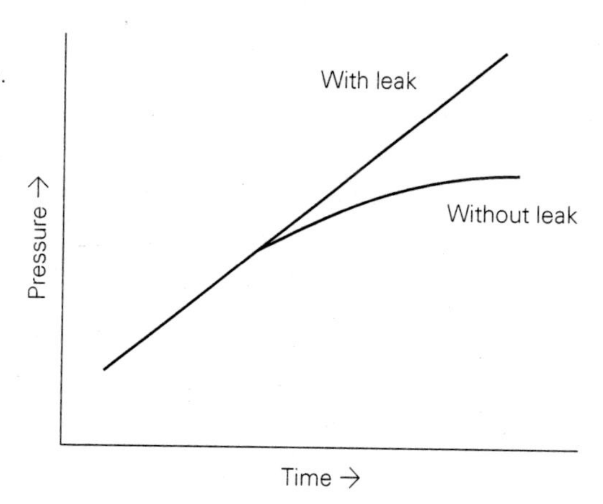

FIGURE 7.1
Diagnostic rate-of-rise curves.

Using these conversions, leaks can be added to the other sources of gas that contribute to the gas load in a vacuum system.

But before we tackle the subject of leak detection, how can we tell whether we have a leak? There is no use spending time and effort looking for a problem that does not exist. As mentioned in chapter 3, the first line of defense is knowing your vacuum system. Comparing pump-down curves is the place to start. A slower pump-down time and a higher-than-normal base pressure for a given pumping time may be indicators of a leak, but they are not the "smoking gun." A rate-of-rise measurement can yield additional data that will help you decide whether to implement a leak detection process or not.

A rate-of-rise measurement is simple to implement and provides a "go/no-go" indicator for leak detection. To perform a rate-of-rise measurement, the system is first pumped down to a preselected pressure. Then the high-vacuum valve is closed, effectively taking the pumping system out of play, and the chamber pressure is monitored as a function of time. If the pressure-versus-time graph shows that the rate of pressure rise increases steadily toward atmospheric pressure over time, then there is a leak and it is time to go into leak detection mode. On the other hand, if the pressure-versus-time graph begins to slowly level off, then the pressure rise is not being caused by a leak but by an internal gas load under the broad heading of outgassing or by diffusion through O-ring seals. Figure 7.1 shows two pressure-versus-time curves, one for outgassing plus a leak and the other for outgassing alone.

7.2 LEAK DETECTION METHODS

There are two standard leak detection methods, known as the inside-out method and the outside-in method. Our example of fixing a bicycle tire is an example of the *inside-out method.* The partially inflated tire tube is submerged in a tub of water. If the tire tube is squeezed to increase the air pressure in the tube, air will be forced out through the hole in the tube. Bubbles will indicate the presence and location of the leak.

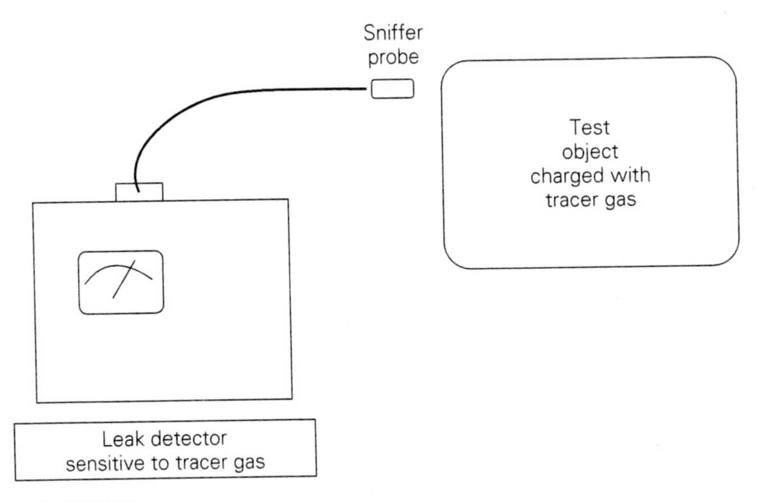

FIGURE 7.2
Inside-out leak detection method.

In most cases, submerging the object to be tested in water is not a good idea. Instead, a tracer gas and leak detector are used. The test object is filled or charged with the tracer gas to a pressure above atmospheric pressure. The leak detector, a mass spectrometer sensitive to the tracer gas, is used to sample, or "sniff," the gas around the test object. If the leak detector detects the tracer gas as the sniffer probe is moved over the exterior surface of the test object, then a leak exists. If no tracer gas is detected, then a leak is not present. Figure 7.2 illustrates the inside-out technique.

The *outside-in method*, on the other hand, applies the tracer gas to the exterior of the test object and uses a leak detector to determine whether any of the tracer gas enters the test object. Figure 7.3 illustrates this method. The test object is connected to the leak detector and is placed under vacuum. Using a spray probe, the tracer gas is applied (sprayed) on the exterior surface of the test object. If a leak is present, some of the tracer gas will travel through the leak into the test object. The leak detector provides a visual or audible signal when it senses the presence of the tracer gas.

With a mass spectrometer leak detector, any gas can be used as the tracer gas. However, our choice can be narrowed if we consider the desired attributes of a tracer gas. Here are a few:

- Nontoxic
- Inert
- Not more than trace quantities present in the atmosphere
- Diffuses readily through minute leaks
- Low absorption rate
- Inexpensive

The first attribute is a safety issue. If the tracer gas is nontoxic, then it won't pose a health threat when sprayed on the test object. This eliminates chlorine and fluorine.

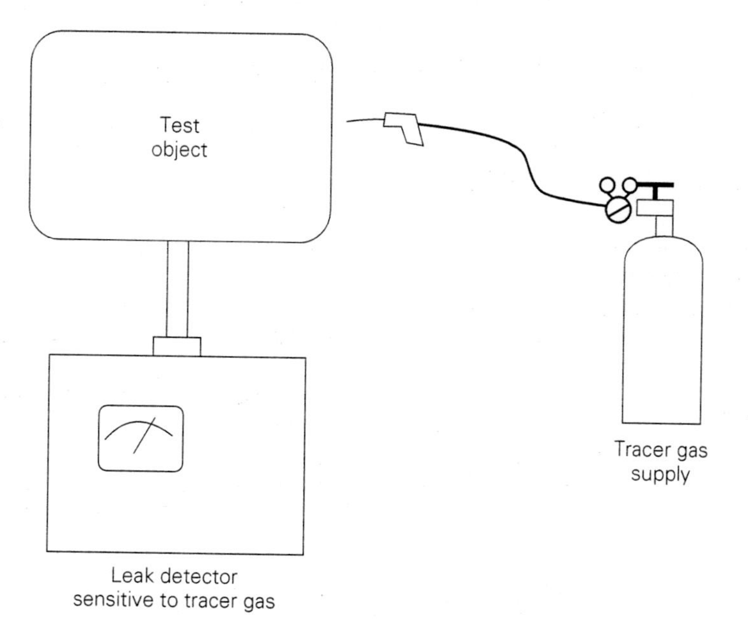

FIGURE 7.3
Outside-in leak detection method.

The second attribute, inert, says that the tracer gas must not react with materials in the vacuum system. The gases in the far-right column of the periodic table are all inert gases, including helium, neon, argon, krypton, xenon.

The third attribute says that the tracer gas must not be common in the atmosphere. Obviously, nitrogen and oxygen are present in too high a concentration to be used for leak detection. The most common inert gas in the atmosphere is argon, 0.94% by volume. Helium is less common at 0.0005% by volume. Nevertheless, there are instances where argon may be used as the tracer gas for leak detection. Neon, krypton, and xenon are even less common in the atmosphere than helium, but are not used for leak detection for other reasons.

The tracer gas must diffuse readily through the leak, minute as it might be. Diffusion of the tracer gas depends on the pressure differential, temperature, and molecular size. Molecular size favors small, light, energetic gas molecules. Only hydrogen, mass $H_2 = 2$, is lighter than helium, mass He = 4, the lightest of the inert gases.

Another attribute of a tracer gas is its low absorption rate. That is, a tracer gas must not absorb into materials in the vacuum system, such as O-ring seals, and thus must persist in the system for a long period of time. Helium, being a monatomic gas, has a very low absorption rate.

And finally, from a financial point of view, the tracer gas should be relatively inexpensive. Helium is less expensive than neon, the next-heavier inert gas.

When all the attributes of an ideal tracer gas are considered, *helium* is by far the best choice. It is nontoxic, inert, not present in the atmosphere in large quantities, diffuses readily through leaks, is slow to absorb, and inexpensive.

7.3 HELIUM MASS SPECTROMETER LEAK DETECTOR

Helium mass spectrometer leak detectors are commonly used in industry to find leaks in vacuum systems. The leak detector consists of a spectrometer tube capable of quantitatively measuring the presence of helium, a vacuum system to maintain an adequately low operating pressure, and sensing and control circuitry.

The heart of a leak detector is the *spectrometer tube*. A cross-sectional view of a spectrometer tube is shown in Figure 7.4. Gas enters the spectrometer tube from the left in the diagram, travels around a series of baffles, and enters the ion chamber. In the ion chamber, gas molecules are ionized and accelerated toward the mass separator, an area of strong magnetic fields. In the mass separator, the trajectory of the ions is changed. The lighter ions are deflected greater than 90 degrees, while the heavier ions are deflected less than 90 degrees. Only the ionized helium atoms are deflected 90 degrees and are able to pass through the slit into the ion collector. The detector in the ion collector creates a current proportional to the partial pressure of helium in the spectrometer tube. The collector current is measured by an electrometer amplifier and displayed on the readout on the control panel.

A conventional leak detector is shown in Figure 7.5. The gas sample enters the test port and flows through a cold trap to remove water vapor and other condensible gases.

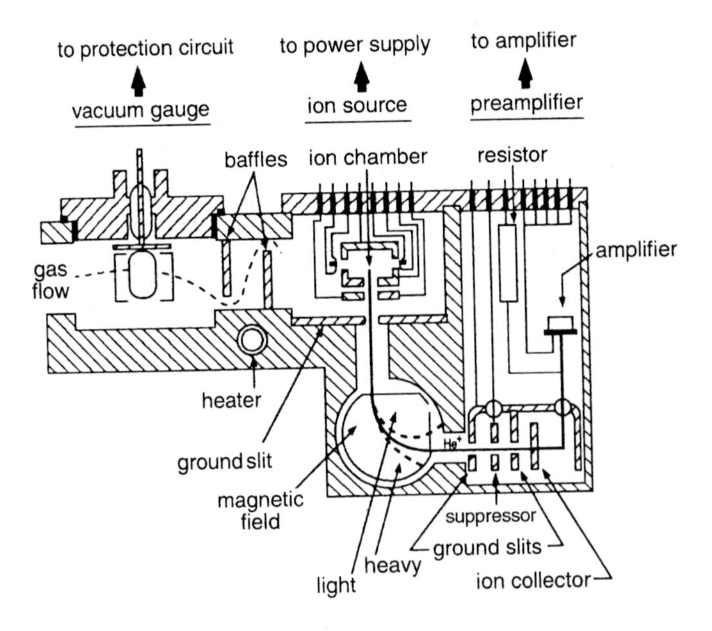

FIGURE 7.4
Spectrometer tube in a leak detector.
Source: Varian Vacuum Products Catalog 2000, p. 231.

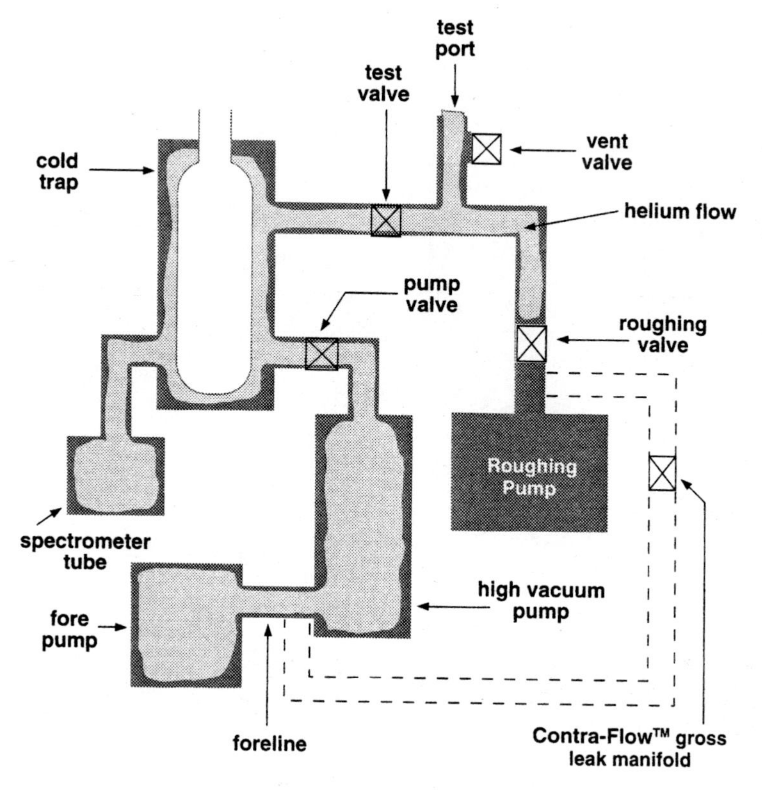

FIGURE 7.5
Conventional helium leak detector.
Source: Varian Product Catalog 2000, Fig. 4, p. 232.

The cold trap is maintained at a pressure (below 10^{-4} torr) in the high-vacuum regime by a high-vacuum pump/forepump combination. The spectrometer tube is also connected to the cold trap and will detect any helium that flows into the cold trap. The advantages of conventional leak detector design are its high sensitivity to helium, which has a low detection limit, and its short response time due to the high volume flow rate at the test port inlet.

An alternative leak detector design uses the *contra-flow method* shown in Figure 7.6. With this method, the system or object to be tested is not connected to the high-vacuum section of the leak detector. Instead, a connection is made to the foreline between the turbomolecular pump and the mechanical backing pump so that the entire gas flow, especially water vapor, does not contribute to the pressure increase in the mass spectrometer. Thus, the contra-flow method does not require a cold trap.

In the contra-flow method, helium must flow against the pumping action of the turbomolecular pump to the mass spectrometer. Helium can do this because it is a energetic, high-velocity gas particle.

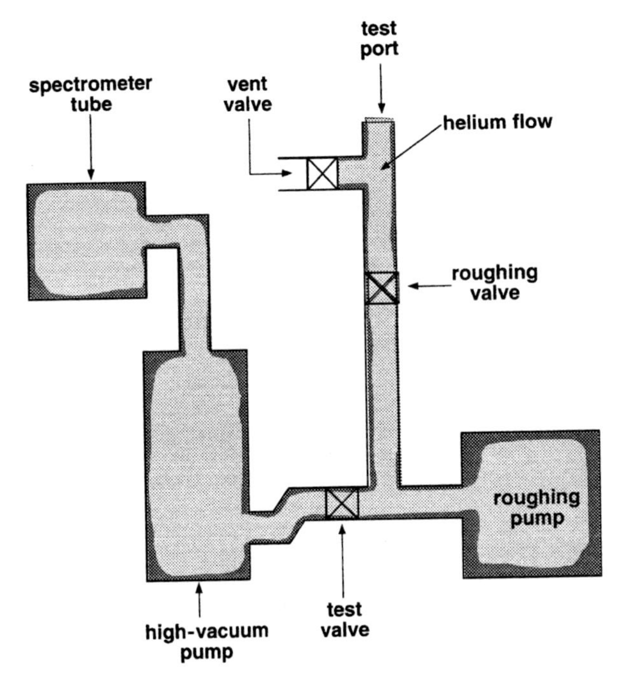

FIGURE 7.6
Contra-flow helium leak detector.
Source: Varian Product Catalog 2000, Fig. 5, p. 232.

The advantages of the contra-flow method are the high permissible inlet pressures and the ability to begin leak detection at a higher pressure. Also, no cold trap is required, thereby reducing the complexity of the leak detector. Both of these advantages make the contra-flow design especially suitable for portable leak detectors.

7.4 CALIBRATION AND STANDARD LEAKS

To obtain quantitatively correct leak rate readings, the sensitivity of the leak detector must periodically be adjusted and calibrated. This requires the use of calibrated leaks. *Calibrated leaks* are commercially available in two types, glass permeation and capillary. The use of these two types of helium calibrated leaks is described in an application note published by Pfeiffer Vacuum entitled "A General Overview of Calibration of Helium Mass Spectrometer Leak Detectors."

A *glass permeation calibrated leak standard* consists of a permeable glass membrane enclosed in a sealed reservoir charged or pressurized with helium (see Figure 7.7). The helium permeates through the glass membrane and then flows through the valve into the leak detector system. The glass membrane is very fragile, so this type of calibrated leak should be handled with care.

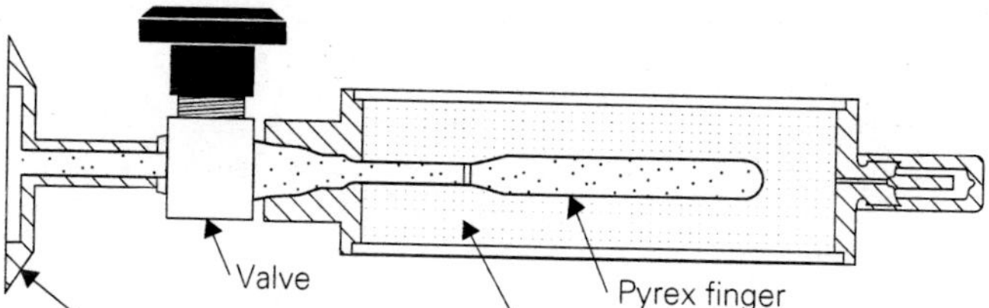

Valve
Pyrex finger
Test port adapter
Helium-filled reservoir

FIGURE 7.7
Glass permeation calibrated leak standard.
Source: Pfeiffer Application Note, Fig. 2.

Since permeation of a gas is temperature dependent, temperature affects the accuracy of the calibrated leak. Temperature coefficients of 3% or more per degree Celsius are common. As the operating temperature of the calibrated leak increases, the leakage rate also increases. Consider the following example.

EXAMPLE 7.1 Given a 2.0×10^{-7}sccs helium, permeation-type leak calibrated at 22°C, what is the leak value at 27°C?
Assume a deviation of 3% per °C.

Solution
The change in temperature, ΔT, is

$$\Delta T = 27°C - 22°C$$

$$\Delta T = 5°C$$

The deviation in percent over 5°C is

$$\text{Deviation} = 0.03/°C \times 5°C \times 100\%$$

$$\text{Deviation} = 15\%$$

The leak rate at 27°C is

$$\text{Leak Rate}_{27°C} = 2.0 \times 10^{-7}\text{ sccs} + (0.15 \times 2.0 \times 10^{-7}\text{sccs})$$

$$\text{Leak Rate}_{27°C} = 2.0 \times 10^{-7}\text{ sccs} + 3.0 \times 10^{-8}\text{ sccs}$$

$$\text{Leak Rate}_{27°C} = 2.30 \times 10^{-7}\text{sccs}$$

A *capillary-type calibrated leak standard* consists of a thin metal or glass capillary tube that extends into a pressurized metal envelope (see Figure 7-8). The gas from the envelope flows through the capillary tube and out through the valve to the leak detector. One advantage of the capillary-type calibrated leak over the permeation-type leak is that it is less sensitive to temperature variations. The leakage temperature coefficient for a capillary-type

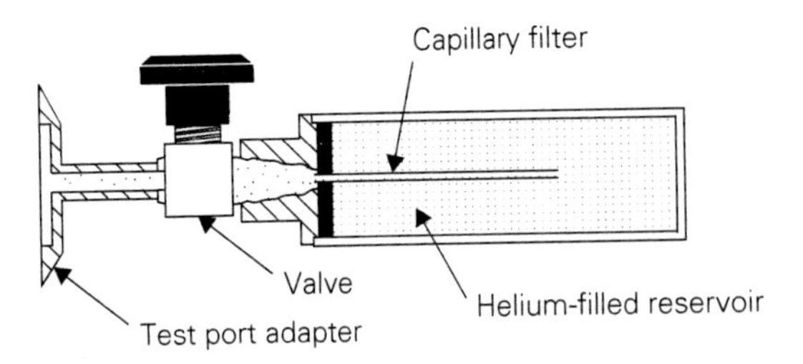

FIGURE 7.8
Capillary-type calibrated leak standard.
Source: Pfeiffer Application Note, Fig. 3.

leak is approximately 0.2% per degree Celsius. Although capillary leaks are not as fragile as permeation leaks, they are susceptible to being plugged by solids and condensation of vapors.

When using a capillary calibrated leak, the leak rate at a given temperature can be found using the same process as illustrated in Example 7.1.

EXAMPLE 7.2

Given a 2×10^{-7}sccs helium, capillary-type leak calibrated at 22°C, what is the actual leak value at 27°C?
Assume a deviation of 0.2% per °C.

Solution

The change in temperature, ΔT, is

$$\Delta T = 27°\text{C} - 22°\text{C}$$
$$\Delta T = 5°\text{C}$$

The deviation in percent over 5°C is

$$\text{Deviation} = 0.002/°\text{C} \times 5°\text{C} \times 100\%$$
$$\text{Deviation} = 1.0\%$$

The leak rate at 27°C is

$$\text{Leak Rate}_{27°\text{C}} = 2.0 \times 10^{-7}\text{ sccs} + (0.010 \times 2.0 \times 10^{-7}\text{ sccs})$$
$$\text{Leak Rate}_{27°\text{C}} = 2.0 \times 10^{-7}\text{ sccs} + 2.0 \times 10^{-9}\text{ sccs}$$
$$\text{Leak Rate}_{27°\text{C}} = 2.02 \times 10^{-7}\text{ sccs}$$

The point of Examples 7.1 and 7.2 is not to show that one type of calibrated leak is better than the other, but to demonstrate that temperature must be taken in consideration so that the exact leak rate is known. If the exact leak rate is known, then it can be used to calibrate the leak detector.

The leak rate of calibrated leaks is also dependent on the age of the device. Calibrated leaks should be stored with their valves open to eliminate gas buildup within the device that can cause inaccuracies. Hence, over time helium gas is constantly escaping from the calibrated leak. This adds up over time. A 10^{-7} or 10^{-8} sccs calibrated leak will lose 2% to 3% of its initial leak rate value over the course of a year. To keep up with internal or external quality standards, calibrated leaks are typically recertified annually.

EXAMPLE 7.3 A helium, permeation-type calibrated leak was last calibrated 3.0 years ago. The label indicates that the leak rate at that time was 2.5×10^{-7} sccs. Assuming a depletion rate of 3% per year, what is the actual value of the leak today?

Solution

The time period, Δt, is

$$\Delta t = 3.0 \text{ years}$$

The depletion rate over 3.0 years is

$$\text{Depletion Rate} = 0.03/\text{yr} \times 3.0 \text{ yr} \times 100\%$$

$$\text{Depletion Rate} = 9.0\%$$

The leak rate at 3.5 years after the last calibration is

$$\text{Leak Rate}_{3.5 \text{ yrs}} = 2.5 \times 10^{-7} \text{ sccs} - (0.09 \times 2.5 \times 10^{-7} \text{sccs})$$

$$\text{Leak Rate}_{3.5 \text{ yrs}} = 2.0 \times 10^{-7} \text{sccs} + 22.5 \times 10^{-9} \text{sccs}$$

$$\text{Leak Rate}_{3.5 \text{ yrs}} \approx 1.8 \times 10^{-7} \text{sccs}$$

Calibrated leaks, traceable to NIST or other standards organizations, like the German Calibration Service controlled by the Federal Institute of Physics and Technology, are commercially available and range from 10^{-5} to 10^{-10} std cc/sec. As Example 7.3 illustrates, the leak begins to age once the calibrated leak leaves the manufacturing area. Examples 7.1 and 7.2 show the effect of temperature on the value of a calibrated leak. Operation and/or storage at an elevated temperature will accelerate the helium gas loss from the leak. This necessitates recertification of the leak after a certain period of time.

7.5 SELECTING A HELIUM LEAK DETECTOR

Your specific application will determine what type of leak detector you should use. Important considerations include leak rates, cycle time, portability, and testing methodology, just to name a few. Regardless of your application, there are some fundamentals that must be considered when choosing a leak detector. They include response time, recovery time, lowest detectable helium leak rate, helium background concentration, pump-down time and inlet pressure crossover, as well as portability and price.

The leak detector's *response time* is simply the time it takes to sense a leak. However, a number of factors contribute to the overall response time of a leak detector. One factor is the time it takes for the helium tracer gas to enter the system through the leak and travel to the leak detector inlet. The volume of the system affects the travel time. Smaller volumes make for faster travel times, and conversely, larger volumes make for longer travel times.

Another factor is *helium pumping speed*. For a given volume and a given leak, the leak detector with a faster helium pumping speed will have a faster response time. A third factor affecting response time is the *conductance* of the line connecting the chamber, or part to be tested, to the leak detector. This could lengthen the response time if a low-conductance path slows down the movement of the helium from the leak to the leak detector's sensing unit. It is common practice to use as short a line as possible with a diameter matching the diameter at the inlet of the leak detector. And finally, the response time is a function of how quickly the electronics in the detection system can react and produce a display.

The *recovery time* of a leak detector indicates how fast it can return to its desired sensitivity after being exposed to a specific leak rate. The time is directly proportional to the amount of helium to which it was exposed and inversely proportional to the helium pumping speed. In mathematical terms, recovery time can be expressed as

$$T_d = \frac{V}{S} \times 2.3 \times \log \frac{Q}{Q_{\min}}$$

where T_d is the recovery or disappearance time in seconds,
V is the volume in liters,
S is the helium pumping speed in liters/second,
Q is the leak rate in atm.cc/second,
and Q_{min} is the smallest detectable leak in atm.cc/second.

The lowest detectable helium leak rate is theoretically the smallest helium signal the leak detector can sense. It is theoretical because these numbers are given for ideal measurement conditions—no helium background and a clean environment. It is an important consideration especially when using the leak detector in conjunction with an auxiliary pumping package in situations where the leak detector itself cannot handle the gas load. This is illustrated in the following example.

EXAMPLE 7.4

Consider the equipment setup in Figure 7.9. The split flow takes place between the auxiliary pumping subsystem and the leak detector. The pumping speed of the vacuum subsystem consisting of a Roots pump and a rotary vane pump is 25 liters per second, and the pumping speed of the pumps in the leak detector is 1 liter per second. What will the leak detector measure and display if the the leak rate is 1×10^{-8} atm cc/sec for helium?

Solution

The split flow depends on the ratio of the pumping speed of each parallel pathway (the vacuum subsystem and the leak detector). Hence,

$$Q_b = \frac{S_b}{S_a + S_b} Q_t$$

$$Q_b = \frac{1\,\text{L/sec}}{1\,\text{L/sec} + 25\,\text{L/sec}} (1 \times 10^{-8} \text{ atm cc/sec He})$$

$$Q_b = 3.85 \times 10^{-10} \text{ atm cc/sec He}$$

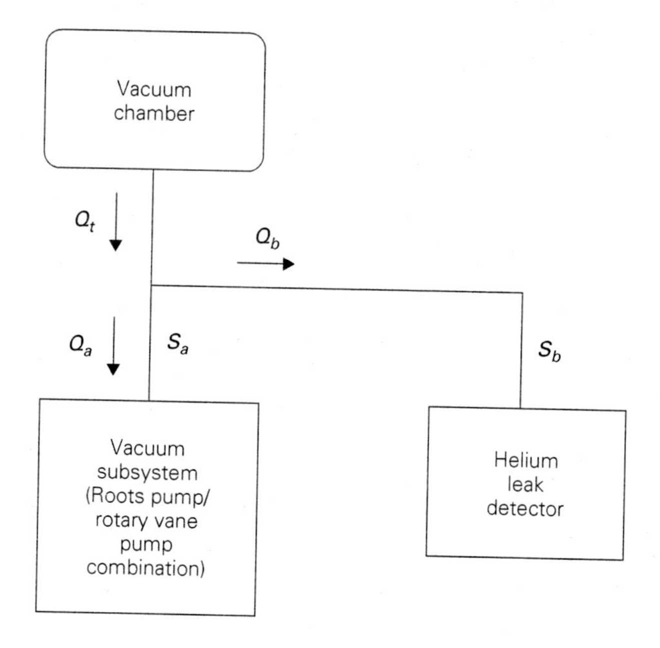

Where: Q_t is the helium leak rate to be detected,

Q_a is the helium pumped by the vacuum subsystem,

Q_b is the helium pumped and measured by the leak detector,

S_a is the helium pumping speed of the vacuum subsystem,

and S_b is the helium pumping speed of the leak detector.

FIGURE 7.9
Figure for Example 7.4.

The helium concentration in air is five parts per million (ppm) and is relatively constant anywhere on earth. This low helium concentration makes helium an ideal tracer gas. An easy solution to correcting for background helium levels is to use the *auto-zero* feature, available on some leak detectors. This effectively tells the electronics to move the reference point up to the helium background level at the time of the test. The operator then sees a leak detector with no background helium contribution while performing the leak test. It is important to realize that it is not necessary to spray a lot of helium when leak testing. Using large amounts of helium is not only a waste, but it will also create problems later by increasing the background helium level around the machine over time.

The *pump-down time* is the time it takes the leak detector's pumping system to reach the inlet test pressure. At that point, the leak detector will go into test mode. This point is also referred to as the *inlet test pressure crossover*. As has been already discussed, pump-down time depends on pumping speed, line conductance, and gas load. In a production environment, higher pumping speeds and inlet test pressure crossovers can reduce cycle times and have a major impact on the production rate at the end of the day.

7.6 USING AN RGA FOR LEAK DETECTION

The use of a *residual gas analyzer* as a leak detector provides some definite advantages over a traditional leak detector system. If the RGA is connected to the process system, it is always available and does not require perturbing the system; stated more simply, the user does not have to roll up a large leak detector cart and attach the leak detector system to the process system. In addition, the process system does not have to be vented to atmospheric pressure.

RGAs can operate in leak detection mode with tracer gases other than helium. For moderate leaks, argon or tetrafluoroethane, a typical gas in cans of aerosol dust remover, can be used as the tracer gas. Only for the smallest leaks must helium be used.

The leak detection process using an RGA is the same as with a traditional leak detector. The RGA must first be placed in the leak detection mode and the mass of the tracer gas must be specified. The partial pressure of the tracer gas is monitored as various joints in the process system are sprayed with the tracer gas. When a leak is sprayed with tracer gas, the partial pressure of the tracer gas will increase or rise. The response is immediate if the leak is the result of a direct path from the outside to the inside of the system.

Figure 7.10 shows the result of a leak test with helium as the tracer gas. The tester moves the helium probe toward and then past the leak, causing the first peak. Once the approximate location of the leak is found, the tester goes back to exactly locate the leak.

There are some situations where conventional leak detectors cannot easily find a leak. For example, leaks across valves that supply gases to a process system cannot be detected with conventional leak detectors, unless, of course, the valve supplies helium.

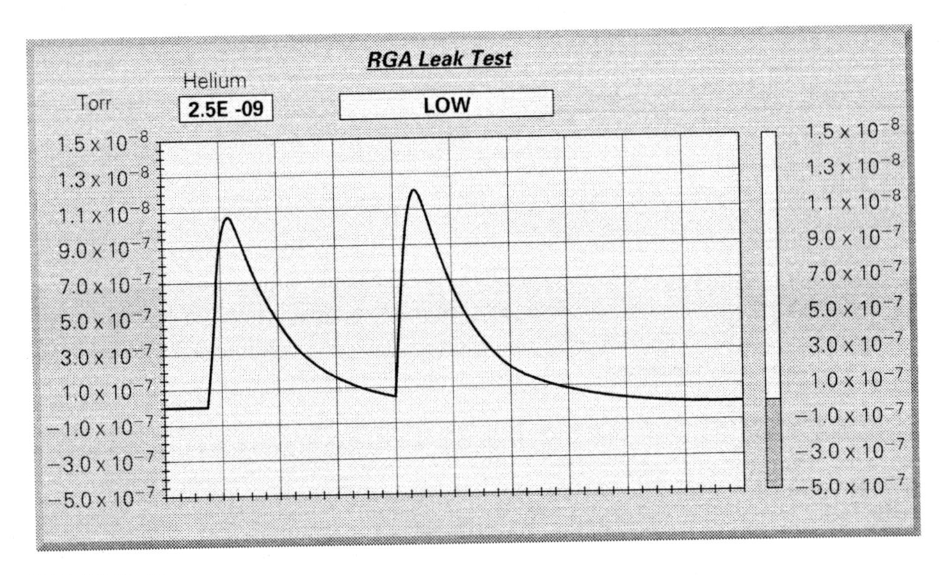

FIGURE 7.10
Helium leak test using an RGA.

With a conventional leak detector, the valve to be tested must be removed from the system and then attached to the leak detector.

Because the RGA can monitor any gas within its operating range, it is unnecessary to remove the valve from the system, and the valves on gas supply lines can be tested *in situ* using a relatively simple procedure. Monitor the partial pressure of the supply gas first with high pressure behind the valve and then with low pressure behind the valve seat. If the partial pressure of the supply gas in question changes, the valve seat is leaking.

Leak-checking supply gas manifolds can also be a challenge. The difficulty is caused by the compression-type fittings used to construct the gas manifold and the large number of connections in close proximity. Compression fittings are difficult to leak-check because the leak is inside the fitting. In this situation, transporting the tracer gas to the leak requires a large flow rate and an extended waiting time for the gas to diffuse into the fitting. Because of the large flow rate and long wait time, it is possible for the tracer gas to travel to adjacent tube fittings and cause a "false positive" of a leak. In these situations, the RGA provides the option of using a tracer gas other than helium. A heavier tracer gas such as argon is far easier to confine to a specific fitting.

SUMMARY

You will inevitably encounter leaks when working with vacuum systems. They don't always occur, and being meticulous in your work will minimize leaks. But when they appear, you will have to use your powers of observation, the logic of a good detective, and some leak detection tools.

The first thing is to determine whether the system behavior is due to a leak or some other problem. If a leak is highly probable, then using a tracer gas and a detector, such as a helium leak detector or an RGA, can help you track down the leak. You can systematically check each section of the system by spraying small amounts of helium and measuring the internal partial pressure of the tracer gas using the leak detector or the RGA. Leak-checking is a skill that needs to be developed, and unfortunately, that often comes only through experience.

BIBLIOGRAPHY

"Application Note 7: Vacuum Diagnostics with a Residual Gas Analyzer." Stanford Research Systems. Sunnyvale, CA.

Deluca, Jean-Pierre. "Choosing a Helium Leak Detector," *Vacuum & Thinfilm*, January 1999, pp. 28–31.

Introduction to Helium Mass Spectrometer Leak Detection, 2nd Edition, Varian Vacuum Products, Lexington, MA, 1995.

Mattox, Donald M. "Leak and Leak Detection," *Vacuum Technology & Coating*, February 2001.

PROBLEMS

1. List the properties that make helium an ideal tracer gas for leak-checking vacuum systems.
2. Why should the connecting lines between a leak checker and the system under test be kept short in length and as large in diameter as possible?
3. What is the difference between a conventional and a contra-flow helium leak detector?
4. Given a 2.5×10^{-7} sccs helium, permeation-type leak calibrated at 22°C, what is the leak value at 20°C? Assume a deviation of 2% per °C.
5. A helium, permeation-type calibrated leak was last calibrated 2.5 years ago. The label indicates that the leak rate at that time was 3.5×10^{-7} sccs. Assuming a depletion rate of 4% per year, what is the actual value of the leak today?
6. Consider the split-flow system shown in Figure 7.9. A different leak detector is used, and its pumping speed is 5 liters per second. What will the leak detector measure and display if the leak rate is 1×10^{-8} atm cc/sec for helium?

LABORATORY ACTIVITY

Rate-of-Rise Test

Equipment Needed: A high-vacuum system and a stopwatch.
Objective: To determine the integrity of a vacuum system using a rate-of-rise test.
Laboratory Procedure: Assume the chamber has been vented to atmosphere.

1. Pump the chamber down to its base pressure.
2. Close the high-vacuum valve and start the stopwatch.
3. Record the time and chamber pressure at 15-second intervals. The sample period can be lengthened if the rate of change in pressure is slow.
4. Take pressure readings until the chamber pressure rises from base pressure to some arbitrary pressure limit such as 1 millitorr.
5. Open the high-vacuum gate valve and pump down the chamber to base pressure.
6. Introduce a small gas leak by opening the gas inlet valve a small amount. For example, open the gas inlet valve until the chamber pressure increases by a factor of 3.
7. Close the high-vacuum valve and monitor the pressure increase.
8. When the upper pressure limit is reached, again open the high-vacuum gate valve and pump down the system.
9. Graph the two pressure curves on the same sheet of semi-log graph paper and note the differences between the two graphs. Repeat the process if the graphs are too similar. In other words, introduce a larger gas leak.
10. When you are finished collecting data, perform the shutdown sequence for the vacuum system.
11. How can rate-of-rise measurements differentiate between a real leak and, say, contamination in the chamber?

CHAPTER

8

Gas Delivery and Pressure Control

8.1 INTRODUCTION

A vacuum system is a part of a larger processing or manufacturing system. Other subsystems might include the gas delivery subsystem and a power delivery subsystem if energy is needed to drive the process being run. A system controller directs each subsystem in order to run the prescribed process recipe and control process parameters. In this chapter, we will focus on delivering process gases and controlling chamber pressure.

In most processing systems, it is necessary not only to pump down the process chamber to some base pressure, but also to introduce process gases and control the chamber pressure at a specified level. This must be done within well-defined control limits.

The control process begins with a precision pressure sensor located on the process chamber such as a capacitance manometer that measures the pressure in the chamber and outputs an electrical signal to the system controller. The system controller compares the signal from the pressure sensor to a desired, or reference, process pressure. If the two signals are within an acceptable tolerance, no adjustment is made. However, if the two signals are not within the acceptable tolerance, a signal is sent to an actuator to adjust the chamber pressure and bring it into closer alignment with the desired process pressure.

The actuator can be either a mass flow controller or a throttle valve. The mass flow controller (MFC) is placed, in terms of gas flow, *upstream* from the process chamber. The

MFC determines how much gas is allowed to enter the process chamber. A throttle valve is placed *downstream* relative to the chamber. In this position in the system, the throttle valve adjusts the conductance in the vacuum line and thus controls how fast gas is being pumped out of the process chamber.

Another method of pressure control suggested in the literature is to vary the pumping speed of the pumps in the vacuum system instead of changing the conductance through the use of a throttle valve. Earlier attempts at implementing this pressure control technique met with limited success, but advances in vacuum pump design have led to the implementation of this technique on some process systems.

Regardless of the pressure control method used, the goal is to achieve a balance between total gas load and exhaust rate in order to achieve the desired operating pressure in the process chamber. Each method will be discussed in greater detail in this chapter, but first we will begin with a brief description of the gas delivery system.

8.2 GAS DELIVERY

Process gases are supplied in gas cylinders as compressed gases or, in some cases, are generated at the point of use, as with nitrogen. Personnel handling compressed gases should be familiar with the hazards before handling the gas. These include chemical hazards as well as hazards associated with high pressures and/or low temperatures. It is also the responsibility of anyone working with gas to know first aid and the properties of the gas being used.

A *compressed gas* is any material or mixture held in a container where the absolute pressure exceeds 40.6 psi or 280 kPa at 68°F (20°C) or any flammable material that is a gas at 68°F (20°C) and 14.7 psia (101.3 kPa). Typically, most compressed gases will not exceed 2,000 to 2,640 psig.

Compressed gases can be classified as nonliquefied or liquefied. A *nonliquefied compressed gas* is a gas, other than a gas in solution under charged pressure, that is entirely gaseous at a temperature of 68°F (20°C). On the other hand, a *liquefied compressed gas,* under the charged pressure, is partially liquid at a temperature of 68°F (20°C). A compressed gas in solution is a nonliquefied compressed gas that is dissolved in a solvent.

Compressed gases can be categorized by their chemical and physical properties as corrosive, cryogenic, flammable, inert, oxidant, and toxic or poisonous. Here is a brief description of these categories of compressed gases taken from the *Design & Safety Handbook for Specialty Gas Delivery Systems* published by Scott Specialty Gases:

- *Corrosive gases.* Gases that corrode material or tissue with which they come in contact, or do so in the presence of water, are classified as corrosive. They are high-pressure, reactive, and can be toxic, flammable, and/or oxidizers. Most are hazardous in low concentrations over long periods of time.
- *Cryogenic gases.* Gases with a boiling point below −130°F (−90°C) at atmospheric temperature are considered cryogenic gases. They are extremely cold and can produce intense burns similar to heat burns, and tissue necrosis may even be more severe. They can be nonflammable, flammable, or oxidizing. Cryogenic liquids

can build up intense pressures. At cryogenic temperatures, system components can become brittle and crack.

- *Flammable gases.* Gases that form a flammable mixture at 13% or less by volume when mixed with air at atmospheric temperature and pressure, or have a flammable range in air of greater than 12% by volume regardless of the lower flammable limit, are classified as flammable. They can be high-pressure, displace oxygen from the air, and can be toxic and reactive. A change in temperature, pressure or oxidant concentration may vary the flammable range considerably.
- *Inert gases.* Gases that do not react with other materials at ordinary temperature and pressure are classified as inert. They are colorless and odorless, as well as nonflammable and nontoxic.
- *Oxidant gases.* Gases that do not burn but will support combustion are classified as oxidants. They can be high-pressure, displace breathing oxygen from air (except O_2 itself), and are toxic and reactive.
- *Toxic or poison gases.* Gases that may chemically produce lethal or other health effects on humans are classified as toxic or poison. They can be high-pressure and reactive, and can be nonflammable, flammable, or oxidizing in addition to their toxicity. The degree of toxicity and the effects will vary depending on the gas; however, death will occur when breathed in sufficient quantities.

During the 1980s, specialty gas-handling systems evolved from simple regulators with pigtail connections to automated high-purity gas manifold systems. This increase in the complexity of gas-delivery systems was driven by three factors: semiconductor process requirements, manufacturing demands, and more stringent government safety regulations.

Conventional specialty gas delivery typically used one gas cabinet for each specialty gas cylinder. A *gas cabinet* such as the one shown in Figure 8.1 is a metal enclosure made of steel with an access door. The gas cabinet provides containment of the gas and offers a means of locally venting any leaks without risk of affecting people. A well-designed gas cabinet can also accommodate manifolds and gas-handing systems to provide precise control over operating parameters and total purging of all equipment and lines.

A properly designed gas cabinet system can fulfill the following objectives:

- Containment of hazardous gas in the event of leakage
- Maintenance of gas integrity
- Automatic shutoff of gas in the event of catastrophic failure
- Effective control of residual gas during cylinder change-out

Using the conventional gas-delivery method, gases can be separated according to their classification. For example, corrosives, oxidizers, flammables and toxic gases can be separated or grouped into separate cabinets. Not only will this satisfy both national and local fire and building codes, but it is also safer in the event of component failure or leakage.

Cabinet exhaust systems should be designed with the capability to change the air in the cabinet many times per minute. For example, a typical gas cabinet designed to allow 150 to 200 linear feet per minute of air to pass through the cabinet when the access window is open provides the equivalent of thirteen air changes per minute.

FIGURE 8.1
Gas cabinet.
Praxair Semiconductor Materials, two-cylinder gas cabinet or Applied Energy Systems, Inc. gas cabinets.

Pressure regulators connected to gas cylinders are used to reduce the pressure of the gas in the cylinder to a workable level that can be safely utilized by operating equipment and instruments. There are two basic types of pressure regulators: single-stage regulators and two-stage regulators.

Single-stage pressure regulators reduce the cylinder pressure to the outlet pressure in one step (see Figure 8.2). Single-stage regulators are good performers for short-duration gas usage and are recommended when precise control of the delivery pressure is not required. Delivery pressure variations will occur with decreasing cylinder pressure.

Two-stage pressure regulators reduce the cylinder pressure to a working level in two steps (see Figure 8.3). The cylinder pressure is reduced by the first stage to a preset intermediate level, which is then fed to the inlet of the second stage. Since the inlet pressure of the second stage is so regulated, the delivery pressure is unaffected by changes in the cylinder pressure.

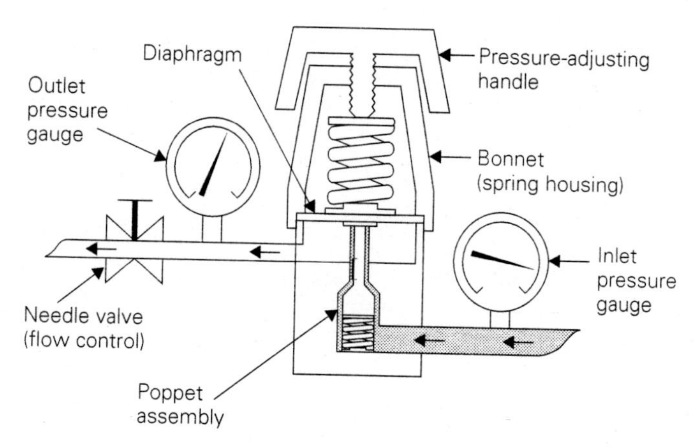

FIGURE 8.2
Single-stage regulator.
Source: Scott Specialty Gases, *Design & Safety Handbook for Specialty Gas Delivery Systems*, p. 10.

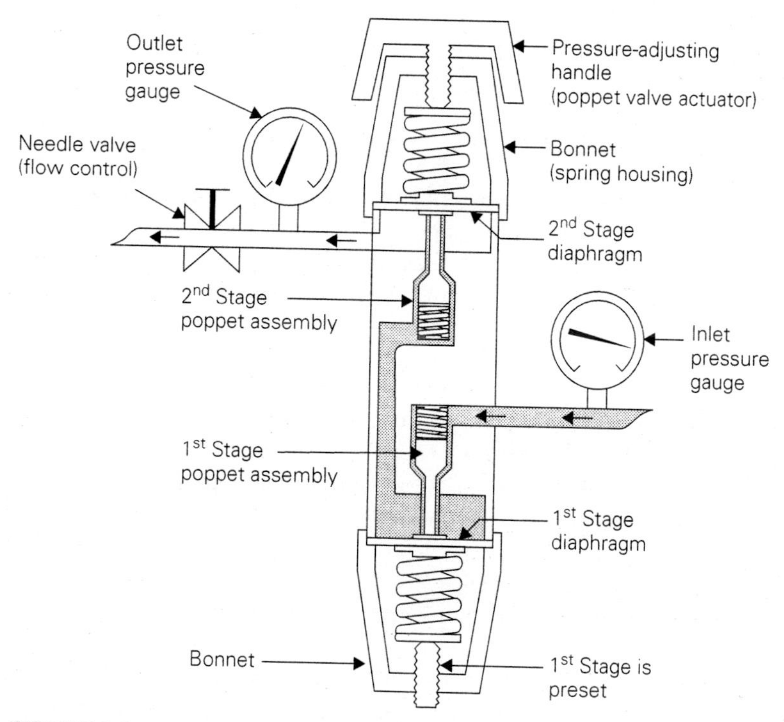

FIGURE 8.3
Two-stage pressure regulator.
Source: Scott Specialty Gases, *Design & Safety Handbook for Specialty Gas Delivery Systems*, p. 10.

In the semiconductor industry and some other manufacturing sectors, conventional gas-delivery systems are giving way to bulk specialty gas systems as wafer fabrication moves to 300 mm wafers. In this manufacturing environment, conventional gas delivery from cylinders requires frequent cylinder changes and multiple connections, which increases the probability of contamination entering the gas stream, decreases the lifetime of gas manifolds, and increases the chances of accidents.

A *bulk specialty gas system* (BSGS), as its name implies, uses large containers to store the gas rather than smaller gas cylinders. These ton containers or trailer bulk sources are located on-site at an appropriate outdoor location. The gas flows from these large tanks into the factory and is distributed to the tools needing the gas.

A *BSGS delivery system* has the potential to provide significant savings through lower raw material pricing and reduced cylinder handling and change-out costs due to larger volume gas storage. However, there is also risk involved, since a BSGS system failure has the potential for process interruption throughout a larger portion of the factory. Appropriate system design engineering and redundancy are required to minimize this risk A BSGS delivery system is shown in Figure 8.4.

Other technical considerations of BSGS systems must also be considered. Bulk systems must operate at much higher flow rates and source pressure and under different thermal conditions than conventional gas delivery systems using gas cylinders. These operating conditions require specialized components, such as regulators, valves, flow controllers, and filters. The biggest challenges are posed by condensable, corrosive specialty gases.

Nevertheless, despite the difficulties encountered when implementing BSGS systems, the advantages can be quite significant. The primary reasons for using BSGS systems are total cost of ownership, safety, product purity, and product consistency.

The cost of ownership of a bulk system offers significant reductions in expenditures for new capital, operations, procurement, startup, system qualification, training, and consumable gas. For example, a ton bulk container can provide the gas-delivery capability of ten standard gas cylinders at roughly one-fifth of the capital cost. Operational savings are achieved through reductions in routine maintenance, labor, parts and materials, and requalification of process lines after gas source changes. However, the major costs savings is in the reduction in the unit cost of the consumable gas, since gas suppliers have the ability to offer lower unit costs owing to the reduction in cylinder preparation, analysis, and rental.

Safety benefits of BSGS systems can be achieved through fewer changeouts, fewer component connections, remote source placement, and less material handling. A significant percentage of gas safety incidents are due to human intervention and many incidents occur during cylinder changeouts.

Product purity and product consistency can be improved through the use of BSGS systems. Impurities can be introduced into the gas system during cylinder changeouts, inadequate purge procedures, or exposure to contaminated equipment. Fewer impurities in the gas system will lead to greater product consistency and higher yields.

FIGURE 8.4
BSGS system.
Picture from BOC website

8.3 MASS FLOW CONTROLLERS

Mass flow controllers (MFCs) are used wherever accurate measurement and control of gas flow is required. Mass flow controllers are usually placed near the process chamber or other point of use.

Mass flow controllers use the thermal properties of a gas to measure the flow rate. That is, the MFC uses the principle that each gas molecule has the ability to pick up heat. This property is called the *specific heat* (C_p) and directly relates to the mass and physical structure of the gas molecule. As shown in Table 8.1, the specific heat is well known for many gases and is generally insensitive to changes in temperature and pressure.

The basic change-in-temperature (Δt) principle is illustrated in Figure 8.5. By adding heat to a gas passing through the sensing tube and monitoring the temperature change, the mass flow can be determined. A heater delivering constant power to the sensing tube is placed between two thermocouple sensing elements.

Mathematically, the heat loss can be described by the following equation:

$$q = \dot{m} C_p \, \Delta T \tag{8.1}$$

where q is the heat lost to the gas flow, $\dot{m}$ is the mass flow, C_p is the specific heat for a constant pressure, and ΔT is the net change in gas temperature as it traverses the tube.

From Equation 8.1, both the specific heat and the flow rate determine the magnitude of the heat flux. Therefore, as the mass and the specific heat vary widely from gas to gas, so does the heat flux. Nevertheless, if the heat flux is monitored, its amplitude being converted to an electrical signal, and the specific heat is known for the gas, the mass flow rate can be determined from the electrical signal.

TABLE 8.1
Specific heat of gases

Gas	Symbol	Specific Heat C_p Cal/g°C	Density G/l @ 0°C
Air	—	0.240	1.293
Ammonia	NH_3	0.492	0.760
Argon	Ar	0.1244	1.782
Arsine	AsH_3	0.1167	3.478
Boron trichloride	BCl_3	0.1279	5.227
Carbon dioxide	CO_2	0.2016	1.964
Carbon monoxide	CO	0.2488	1.250
Chlorine	Cl_2	0.1144	3.163
Diborane	B_2H_6	0.508	1.235
Dichlorosilane	SiH_2Cl_2	0.150	4.506
Fluorine	F_2	0.1873	1.695
Freon-14	CF_4	0.1654	3.926
Freon-23	CHF_3	0.176	3.127
Freon-116	C_2F_6	0.1843	6.157
Helium	He	1.241	0.1786
Hydrogen	H_2	3.419	0.0899
Hydrogen bromide	HBr	0.0861	3.610
Hydrogen chloride	HCl	0.1912	1.627
Hydrogen fluoride	HF	0.3479	0.893
Krypton	Kr	0.0593	3.739
Neon	Ne	0.246	0.900
Nitrogen	N_2	0.2485	1.250
Nitrous oxide	NO	0.2088	1.964
Oxygen	O_2	0.2193	1.427
Phosphine	PH_3	0.2374	1.517
Silane	SiH_4	0.3198	1.433
Silicon tetrachloride	$SiCl_4$	0.1270	7.580
Silicon tetrafluoride	SiF_4	0.1691	4.643
Sulfur dioxide	SiO_2	0.1488	2.858
Sulfur hexafluoride	SF_6	0.1592	6.516
Thichlorosilane	$SiHCl_3$	0.1380	6.043
Tungsten hexafluoride	WF_6	0.0810	13.28
Xenon	Xe	0.0378	5.858

As shown in Figure 8.5, a small portion of the flow is shunted through a smaller sensor tube. The percentage of the mass flow through the sensor relative to the mass flow through the bypass is constant over the operational range of the device.

When there is no gas flow through the sensor tube, the two thermocouple sensors will measure the same temperature, and the voltage difference between the two thermocouple outputs will be zero. On the other hand, when gas does flow through the sensor tube, the gas is warmed by the heater, and the downstream thermocouple detects a higher temperature than the upstream thermocouple. This will produce a voltage difference

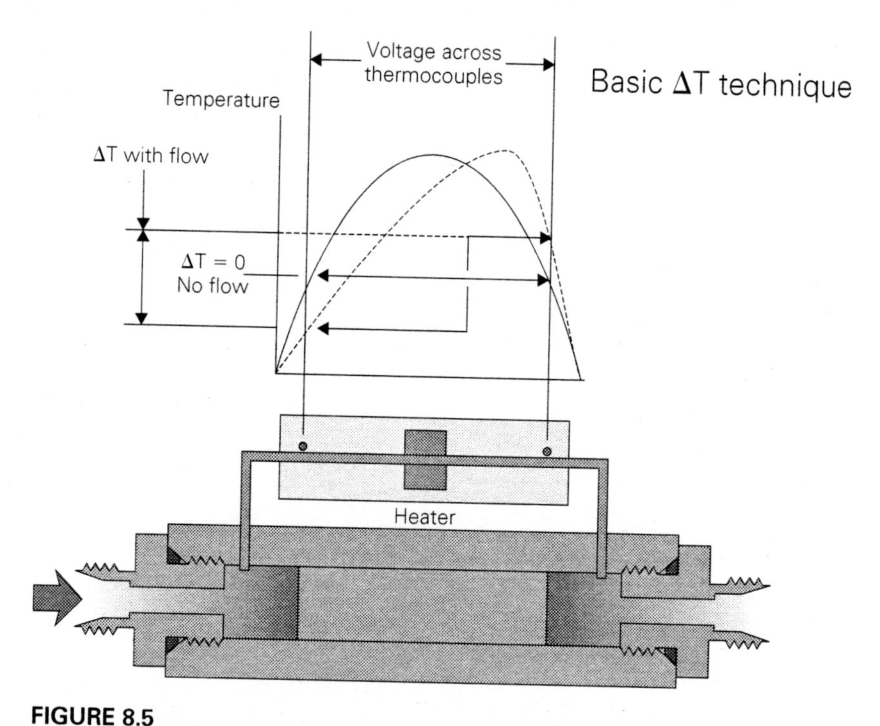

FIGURE 8.5
Basic ΔT technique for measuring mass flow of a gas.
Source: MKS, *Introduction to Mass Flow Controllers,* p. 9.

between the outputs of the two thermocouples, and this voltage difference can be correlated with the mass flow rate.

An improvement to the design of the temperature-sensing circuit is to replace the thermocouple sensors with *resistance temperature detectors,* or RTDs, which measure the temperature of the gas (see Figure 8.6). A constant current source powers the coils. The current is converted to heat by the resistance of the wire. Since the resistance of the coils varies with temperature, the coils can function as resistance temperature detectors.

Two RTDs form two sides of a Wheatstone Bridge circuit. When there is no gas flow through the sensor, the legs of the Wheatstone Bridge are balanced, and the voltmeter reads zero. As flow increases, the temperature difference between the two RTDs increases, unbalancing the Wheatstone Bridge and producing a nonzero voltage reading on the voltmeter. The control circuitry amplifies the output voltage, linearizes it, corrects for gas type and flow range, and finally produces an output voltage in the range of 0–5 VDC that corresponds to the mass flow rate.

A major disadvantage of using RTDs is their slow response time. Every time the gas flow changes, the entire sensor assembly has to shift to a new temperature profile. This may take many seconds to minutes, which may be too long for process control applications that require a fast response.

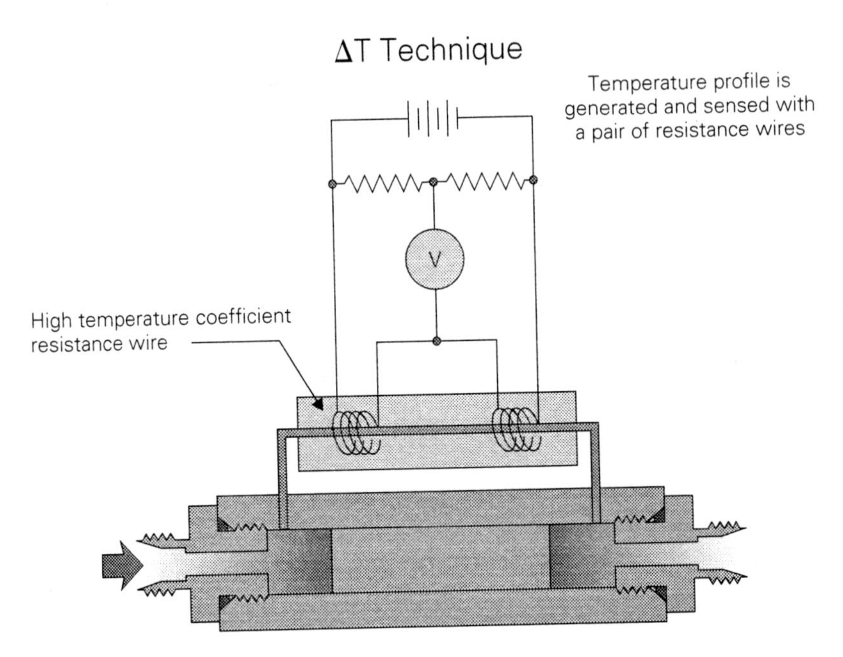

FIGURE 8.6
Mass flow measurement using RTDs.
Source: MKS, *Introduction to Mass Flow Controllers,* p. 10.

The slow response of the RTD method can be overcome by implementing a constant temperature scheme, similar to that used in the Pirani pressure gauge (see Figure 8.7). In this technique, power to the heaters is varied as the flow conditions change to keep the temperature profile constant. The result is response times on the order of tens of milliseconds rather than tens of minutes.

Many MFCs, like the one shown in Figure 8.8, employ a *PID control system* to maintain a constant gas flow. PID stands for "proportional-integral-derivative." As its name implies, the control system looks at three attributes of the input signal, in this case the error signal, and generates an output or control signal that is a function of these three attributes. In mathematical terms,

$$\text{Output} = K_p \times \text{Error} + K_i \int (Error)\, dt + K_d \frac{d\ \text{Error}}{dt} \tag{8.2}$$

where K_p, K_i, and K_d are weighting or gain factors.
The output signal is the input to the valve driver that controls the position of the proportioning control valve.

Let's look at the output signal equation in a little more detail. The first term is the proportional term. Its contribution to the output signal is proportional to the error signal as determined by the weighting factor, K_p. This terms contribution to the output changes as fast or as slow as the error signal.

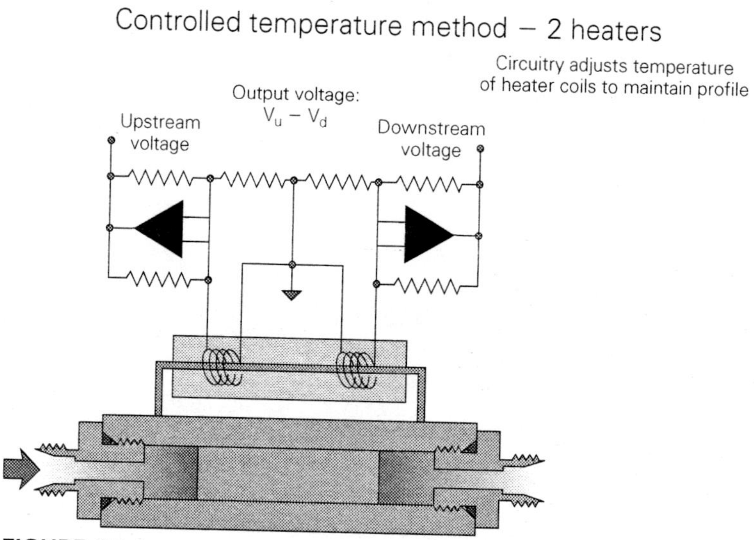

Controlled temperature method – 2 heaters

Circuitry adjusts temperature of heater coils to maintain profile

Output voltage: $V_u - V_d$

Upstream voltage

Downstream voltage

FIGURE 8.7
Controlled temperature method to measure gas flow.
Source: MKS, *Introduction to Mass Flow Controllers,* p. 11.

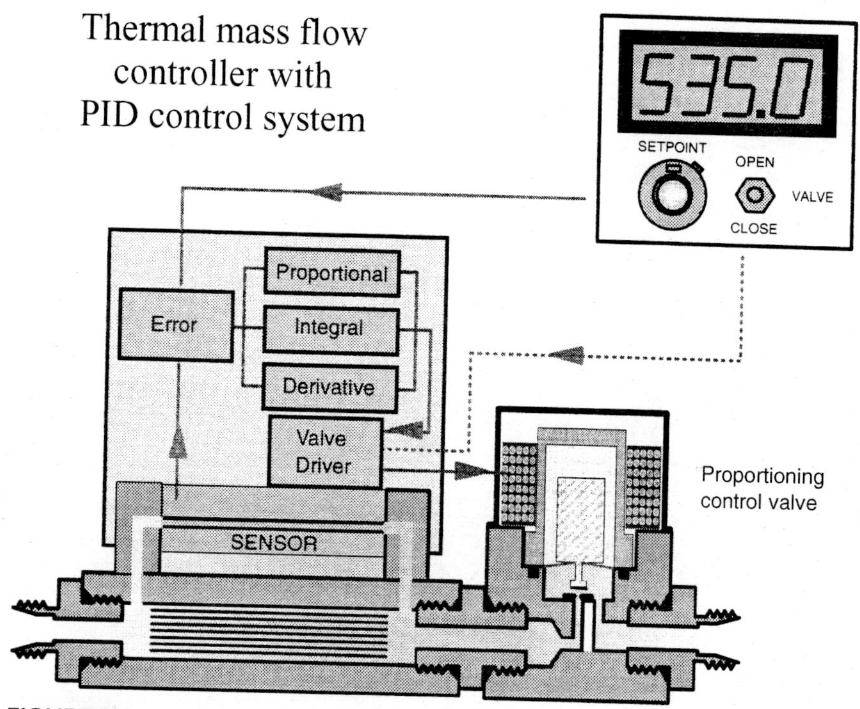

FIGURE 8.8
MFC with PID control.
Source: MKS, *Introduction to Mass Flow Controllers,* p. 13.

The second term is the integral term. The integral of a function is the area under the function curve. Hence, the $K_i\int(Error)dt$ term is proportional to the area under the error curve. This contribution to the output signal corrects for slow-changing DC offsets from the desired level. That is, a small DC offset will accumulate over time and the integral term will grow, eventually causing the control system to correct for the DC offset.

The last term is the derivative term. The derivative term looks at the rate of change, or slope, of the error signal. If the error signal is not changing, the rate of change is zero and the contribution of the derivative term is zero. However, if the error signal is changing, the derivative has a nonzero value, either positive or negative depending on the direction of change in the signal. If the error signal is changing very rapidly, this term's contribution to the output signal can be large; if the error signal is changing slowly, the contribution of this term to the output signal will be small.

The response of a PID control system to a change in the pressure in the process chamber can fall into one of three categories: (1) over-damped response, (2) critically-damped response, and (3) under-damped response. To illustrate these three responses, let's assume that a step change occurs in the chamber pressure—that is, a new steady-state value.

In an over-damped response, as the step-change in the chamber pressure is initiated, the output rises slowly and, after a relatively long time, reaches the final (steady-state) value. The output never overshoots the final value and slowly approaches it.

A critically-damped response is very similar to an over-damped response. However, the time needed to reach the final steady-state value is shorter than in the over-damped case, and without any oscillation or overshoot. In many applications, this is the ideal case or most desirable response.

The under-damped response exhibits a very fast rise to the steady-state value. However, the response exhibits overshoot, where the pressure rises above the desired steady-state value. The control system then corrects the pressure. This results in undershoot. Ultimately, a series of oscillations of decreasing amplitude ensues. This type of response results in the shortest time to reach the desired steady-state value, but the settling time will be longer than for the critically-damped case. The three possible PID responses are illustrated in Figure 8.9.

Figure 8.10 shows a graph illustrating the operation of an MFC responding to a change in set point where the final set point is higher than the initial set point. Some ringing indicates that the system is slightly under-damped and results in some transient overshoot and transient undershoot. The settling time is the time it takes for the MFC to reach a final steady-state value within 2% of the final set point.

Mass flow controllers must be calibrated to the type of gas flowing through the unit, because gas flow is measured indirectly. Relevant gas properties include specific heat, gas density, and molecular structure. Since most MFCs are calibrated with nitrogen, a *gas correction factor* (GCF) must be used to produce a correct reading as gas flow.

The GCF can be stated in mathematical form as

$$\begin{aligned} \mathit{Gas\ Correction\ Factor} &= \mathit{GCF} \\ &= \frac{0.3106 \times \mathit{Molecular\ structure\ correction\ factor}}{\text{Density at STP (mgms/liter)} \times \text{Specific heat (cal/gm} - {}^\circ\text{C)}} \end{aligned} \quad \textbf{(8.3)}$$

A) Critically-damped response

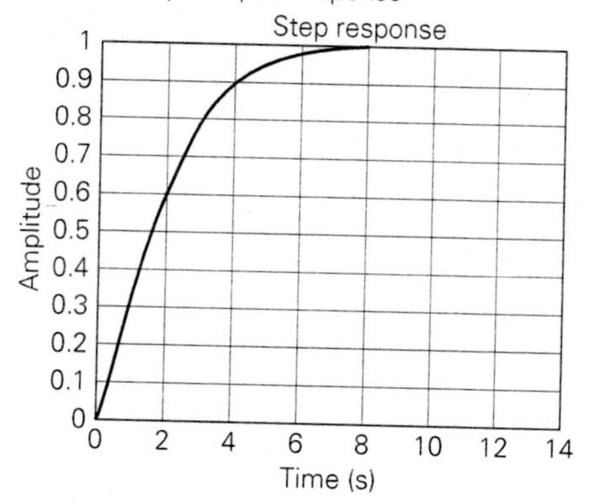

B) Under-damped response [Note: Change y-scale to 0 to 1.6. Waveform settles to 1.0]

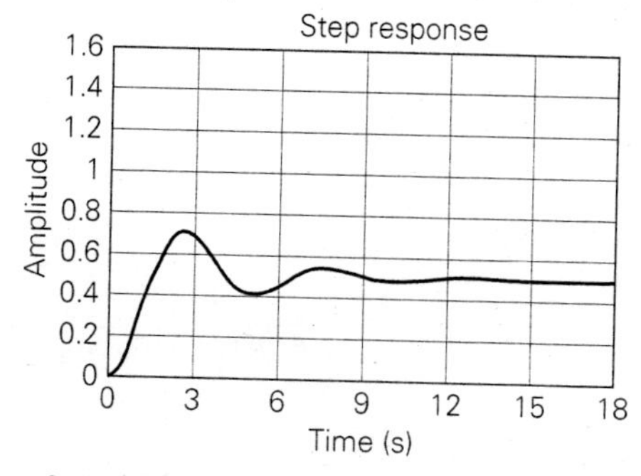

C) Over-damped response

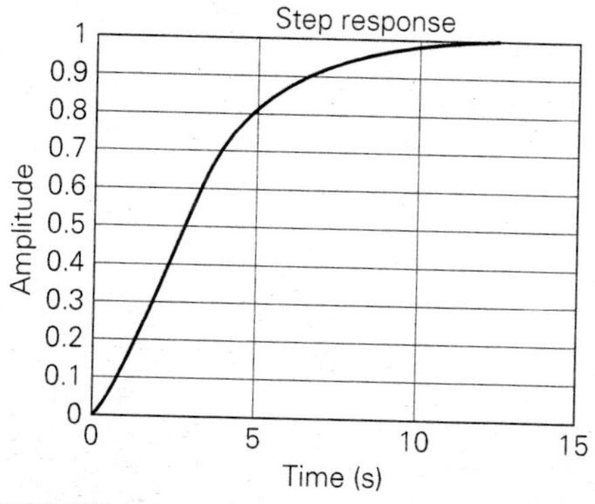

FIGURE 8.9
Over-damped, critically-damped, and under-damped responses in a PID control system.

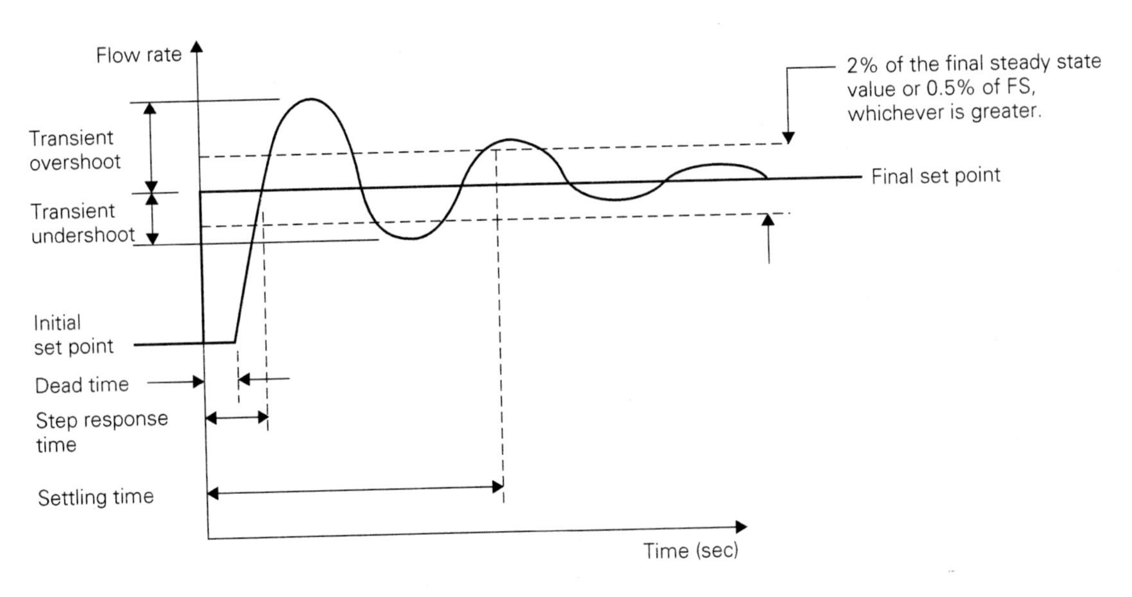

FIGURE 8.10
Time response to a change in set-point.

where 0.3106 is the product of the density of nitrogen at STP[1] and the specific heat of nitrogen. The molecular structure correction factor is 1.030 for monatomic gases, 1.000 for diatomic gases, 0.941 for triatomic gases, and 0.880 for polyatomic gases. Calculated GCFs are not precisely correct under all circumstances, and the flow-reading accuracy may vary a few percent. Precise calibration requires the use of a primary standard with the actual gas. However, this will not affect repeatability.

Table 8.2 provides a list of GCFs for commonly used gases. Note that nitrogen has a GCF of 1.00, as does air, which is 80% nitrogen. All other gases have GCFs either larger or smaller than 1.00. To determine the flow of a specific gas, multiply the MFC reading times the GCF for the gas flowing through the MFC. That is,

$$Q_x = GCF \times Q_{MFC\ Readout} \tag{8.4}$$

where Q_x is the flow rate of the actual gas flowing through the MFC.

EXAMPLE 8.1 Using an MFC, argon gas is introduced into a process chamber. Determine the argon flow rate if the MFC readout indicates a reading of 75 sccm. Assume that the MFC is calibrated for nitrogen gas.

Solution
Using Equation 8.4

$$Q_x = GCF \times Q_{MFC\ \text{Readout}}$$

[1]From Table 8.2, the specific heat for nitrogen is 0.2485 and the density is given as 1.250. Multiplying the two values together yields 0.3106.

and the GCF for argon, 1.45, we obtain the following expression:

$$Q_{Ar} = 1.45 \times 50 \text{ sccm}$$
$$Q_{Ar} = 72.5 \text{ sccm}$$

EXAMPLE 8.2 Determine the MFC reading for an SiH4 flow rate of 75 sccm. Assume that the MFC is calibrated for nitrogen.

Solution

In this case, we will need to solve Equation 8.4 for the MFC readout flow rate. Doing this yields

$$Q_{MFC\ \text{Readout}} = \frac{Q_{SiH_4}}{GCF_{SiH_4}}$$

Substituting the desired SiH_4 flow rate and corresponding GCF yields

$$Q_{MFC\ \text{Readout}} = \frac{75 \text{ sccm}}{0.6}$$
$$Q_{MFC\ \text{Readout}} = 125 \text{ sccm}$$

There are many different mass flow controllers on the market. For example, the MKS-1179 is a general-purpose mass flow controller designed to measure and control the flow of gases in a wide variety of applications. Technical data for the MKS-1179 is provided in Table 8.3.

TABLE 8.2
Gas correction factors.

Gas	Symbol	Specific heat C_p Cal/g°C	Density G/I @ 0°C	Conversion factor
Air	—	0.240	1.293	1.00
Ammonia	NH_3	0.492	0.760	0.73
Argon	Ar	0.1244	1.782	1.39
Arsine	AsH_3	0.1167	3.478	0.67
Boron trichloride	BCl_3	0.1279	5.227	0.41
Carbon dioxide	CO_2	0.2016	1.964	0.70*
Carbon monoxide	CO	0.2488	1.250	1.00
Chlorine	Cl_2	0.1144	3.163	0.86
Diborane	B_2H_6	0.508	1.235	0.44
Dichlorosilane	SiH_2Cl_2	0.150	4.506	0.40
Fluorine	F_2	0.1873	1.695	0.98
Freon-14	CF_4	0.1654	3.926	0.42
Freon-23	CHF_3	0.176	3.127	0.50
Freon-116	C_2F_6	0.1843	6.157	0.24
Helium	He	1.241	0.1786	1.45

(Continued)

TABLE 8.2 ***(Continued)***
Gas correction factors.

Gas	Symbol	Specific heat C_p Cal/g°C	Density G/l @ 0°C	Conversion factor
Hydrogen	H_2	3.419	0.0899	1.01
Hydrogen bromide	HBr	0.0861	3.610	1.00
Hydrogen chloride	HCl	0.1912	1.627	1.00
Hydrogen fluoride	HF	0.3479	0.893	1.00
Krypton	Kr	0.0593	3.739	1.543
Neon	Ne	0.246	0.900	1.46
Nitrogen	N_2	0.2485	1.250	1.00
Nitrous oxide	NO	0.2088	1.964	0.71
Oxygen	O_2	0.2193	1.427	0.993
Phosphine	PH_3	0.2374	1.517	0.76
Silane	SiH_4	0.3198	1.433	0.60
Silicon tetrachloride	$SiCl_4$	0.1270	7.580	0.28
Silicon tetrafluoride	SiF_4	0.1691	4.643	0.35
Sulfur dioxide	SiO_2	0.1488	2.858	0.69
Sulfur hexafluoride	SF_6	0.1592	6.516	0.26
Thichlorosilane	$SiHCl_3$	0.1380	6.043	0.33
Tungsten hexafluoride	WF_6	0.0810	13.28	0.25
Xenon	Xe	0.0378	5.858	1.32

TABLE 8.3
Technical data for MKS-1179 mass flow controller.

Full-scale ranges (N_2 Equivalent)	10, 20, 50, 100 sccm
Maximum inlet pressure	150 psig
Normal operating pressure differential (w/atm pressure at MFC outlet)	
10 – 5000 sccm	10 to 40 psi differential
10000 – 30000 sccm	15 to 40 psi differential
Control range	2% to 100% of full-scale (F.S.)
Accuracy	± 1.0% of full-scale
Controller settling time	< 2 seconds (to within 2% of set point)
Repeatability	± 0.2% of full-scale
Resolution	0.1% of full-scale
Temperature coefficients	
Zero	< 0.04% of F.S./°C
Span	< 0.08% of reading/°C
Connector types	15-pin Type "D", 20-pin card edge
Leak integrity	
External (scc/sec. He)	$< 1 \times 10^{-9}$
Through closed valve	<1.0% of F.S. at 40 psig inlet to atm. (To assure no flow-through, a separate positive shut-off valve is recommended.
Power	± 15 VDC
Warm-up	< 5 minutes

8.4 DOWNSTREAM PRESSURE CONTROL

Pressure in a process chamber can be controlled in two basic ways: (1) downstream pressure control, and (2) upstream pressure control. In each method, the main system components include a process chamber, a gas supply, a pressure transducer, and a control element. The terms *downstream* and *upstream* refer to the direction of gas flow, in much the same sense as water flowing in a river. In *downstream pressure control,* the control element, usually a throttle valve, is placed between the process chamber and the vacuum pumping system. On the other hand, in an *upstream pressure control* system, the control element, usually a mass flow controller, is placed between the gas supply and the process chamber.

Downstream control is the method of choice in most semiconductor processes to manage the partial pressure of process gases. Proper partial pressure is achieved by independently controlling the mass flow of input gases while at the same time controlling the total pressure of the gas mixture. Transient increases in the gas load due to reaction by-products can be rapidly removed by increasing the input gas flow that will automatically open the throttle valve, thereby increasing the pumping speed. In addition, cleanup between process steps can be expedited by fully opening the throttle valve while still under control, thus eliminating the need for expensive by-pass valving.

The main components of a downstream pressure control system include a gas supply, a process chamber, a variable conductance valve with controller, a pressure transducer, and a vacuum pumping system. Typically, the variable conductance valve is a throttle valve of the butterfly type constructed of 316 stainless steel for high corrosion resistance. A high-precision stepper motor is used to position the throttle plate through a belt-driven gear-reduction system. Figure 8.11 shows a block diagram of a typical downstream pressure control system.

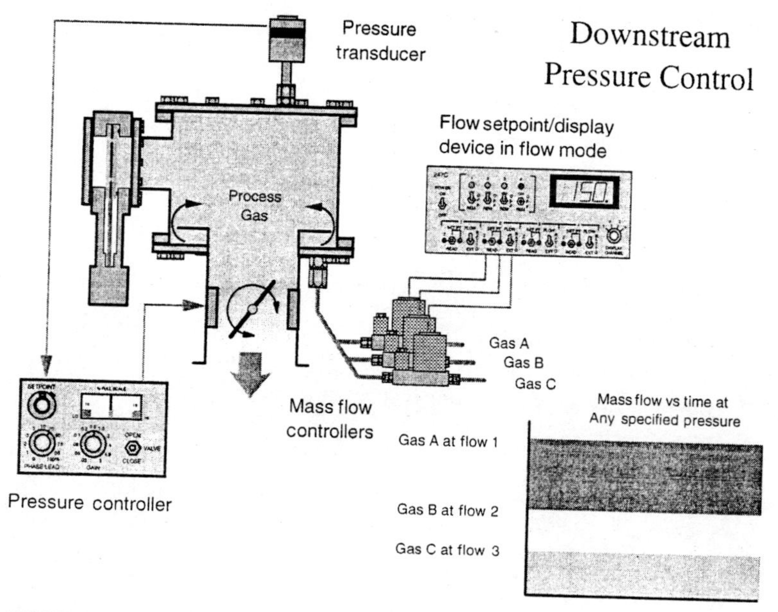

FIGURE 8.11
Downstream pressure control system.
Source: MKS Instruments, *Introduction to the Creation & Control of the Vacuum Process Environment,* p. 83.

The controller is a closed-loop feedback-control system. Gas is admitted to the process chamber through mass flow controllers under the control of the host computer system. After passing through the process chamber, the gas flows from the process chamber through the restriction in the throttle valve and then to the pumping system. To achieve a specific partial pressure, the throttle valve must be driven to an appropriate conductance value. The throttle plate is adjusted by the throttle controller, which compares the pressure measured by the pressure transducer to the desired pressure value dictated by the host computer or the operator from the keypad and produces an error signal. This error signal then drives the stepper motor controlling the valve position, thereby reducing the error signal to zero.

Here is another way of looking at the downstream pressure control process. The chamber pressure P is determined by the pumping speed S under a constant gas flow Q. Mathematically, this can be stated as,

$$P = \frac{Q}{S_{net}} \tag{8.5}$$

Downstream pressure control changes the pumping speed by varying the conductance C of the controlling valve which varies the overall conductance of the vacuum exhaust line. Thus,

$$\frac{1}{S_{net}} = \frac{1}{S_{pump(s)}} + \frac{1}{C}. \tag{8.6}$$

There are some disadvantages to downstream control. For example, using a valve in the process gas flow causes localized perturbations in the gas flow that affect both the velocity of the gas and particle formation. Also, reaction by-products may deposit in the valve. To address this issue, valves must be fully reconditioned during each scheduled preventative maintenance (PM) procedure of a chamber in order to remove or minimize this source of contamination.

8.5 UPSTREAM PRESSURE CONTROL

Upstream pressure control is applied in situations where the controlling element is located between the gas source and the process chamber, "upstream" in terms of the gas flow. This system is designed to control pressure using a single, or possibly a premixed, process gas.

The system components used in an upstream pressure control system are similar to those used in a downstream pressure control system (see Figure 8.12). System components include a gas supply, a process chamber, a pressure transducer, and a vacuum pumping system. The difference is that the control element is usually a mass flow controller.

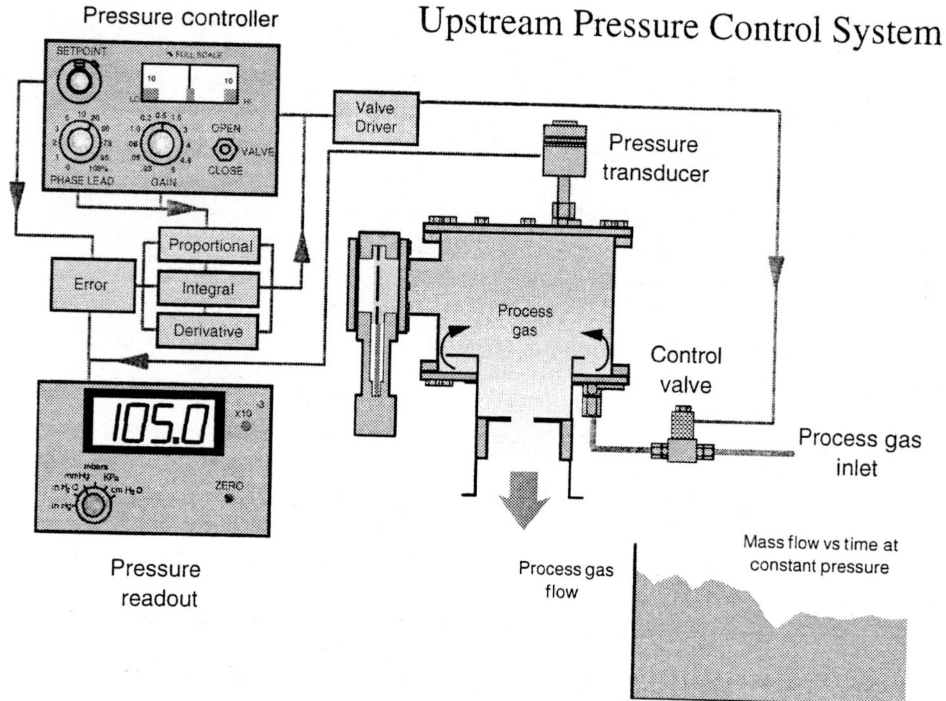

FIGURE 8.12
Upstream pressure control system.
Source: MKS Instruments, *Introduction to the Creation & Control of the Vacuum Process Environment,* p. 81.

SUMMARY

Gas delivery begins in the sub-fab or outside the building in a gas storage area. Inert gases are stored in gas cylinders or large bulk storage, whereas source gases are stored in special gas cabinets equipped with gas sensors that will detect any gas leaks.

The gas flows through a maze of stainless steel piping to a cleanroom above the sub-fab. This gas piping is directed to a gas box located near or on the process tool. In the gas box, a system of mass flow controllers and valves regulate the flow of gas into the process chamber.

Mass flow controllers use the heat-carrying, or thermal, properties of a gas to measure the gas flow rate. The amount of gas flowing through a sensing tube can be determined by adding heat to the gas passing through the tube and monitoring the temperature of the gas at the inlet to the heater and the outlet from the heater. Since the specific heat, or thermal conductivity, of gases differs, mass flow controllers must be calibrated to the specific gas that is flowing through the MFC. Manufacturers of MFCs provide gas calibrating tables of commonly used process gases.

Controlling chamber pressure is critical to a manufacturing process. By using MFCs and valves to control the gas input to the process chamber and throttle valves in the process chamber's exhaust line, the right chamber pressure, and therefore the right amount of process gases, can be achieved. Two common configurations are upstream pressure control and downstream pressure control.

Upstream pressure control uses MFCs to control the gas input to the process chamber. An increase in the throughput of the MFC raises the chamber pressure. Conversely, a decrease in the throughput of the MFC lowers the chamber pressure. In contrast, downstream pressure control uses a throttle valve to increase or decrease the amount of gas being removed from the process chamber. Less conductance in the exhaust line raises the chamber pressure, and conversely, more conductance in the exhaust line lowers the chamber pressure.

BIBLIOGRAPHY

Design & Safety Handbook for Specialty Gas Delivery Systems, Scott Specialty Gases, Plumsteadville, PA, Millenium Edition.

Gupta, Sudhir. *Elements of Control Systems,* Pearson Education, Saddle River, NJ, 2002.

"Introduction to the Creation & Control of the Vacuum Process Environment." MKS Instruments, Wilmington, MA.

PROBLEMS

1. List an advantage and a disadvantage of using thermal mass flow meters to measure gas flow.
2. A mass flow controller calibrated for nitrogen is being used to measure the flow of argon into a process chamber. If the MFC reads 125 sccm, what is the actual argon flow rate?
3. A mass flow controller calibrated for nitrogen is used to measure the flow of SiH_4 into a process chamber. The MFC reads 100 sccm. What is the actual SiH_4 flow rate?
4. If a flow of 200 sccm of argon is required for a process, what would be the equivalent MFC reading? Assume that the MFC is calibrated for nitrogen.
5. If a flow rate of 50 sccm of SiH_4 is required, what would be the equivalent MFC reading? Assume that the MFC is calibrated for nitrogen.

CHAPTER 9

Safety Issues in Vacuum Systems

9.1 INTRODUCTION

Vacuum system safety begins with reading and following safety instructions, applying sensible precautions, and being vigilant for situations that might pose an elevated risk to you and those around you. For our discussion, vacuum system safety will be divided into the following categories:

1. Electrical Hazards
2. Mechanical Hazards
3. Vacuum Hazards
4. Thermal Hazards
5. Pressurized Devices
6. Chemical Hazards

The vacuum subsystem of most processing systems is just one component of a more complex piece of equipment. Other subsystems will also introduce safety issues such as the power delivered by RF or microwave generators and matching units. This chapter will only address safety concerns directly related to vacuum system components; it will not address the safety concerns of other subsystems within a processing system.

9.2 ELECTRICAL HAZARDS

Electric shock is a common risk when using any equipment powered by electricity. There are many situations that could put you at risk of receiving an electric shock. Here are a few examples: improperly connected equipment or instruments, short circuits, exposed conductors, pinched wires, damaged insulation, and ungrounded components.

The physiological responses to electric shock range from a slight tingling sensation to disruption of the electrical activity of the heart and burns to body tissue (see Table 9.1). For DC currents, the perception level is about 5 mA for men and 3 mA for women. The minimum level that we can perceive occurs at the power-line frequency of 60 Hz. At 60 Hz, currents as low as 1 mA can be felt. At frequencies above 60 Hz, the minimum current that can be perceived increases and is approximately 3 mA at 1 kHz and 12 mA at 10 kHz.

As the current level increases above the perception level, a person retains enough voluntarily control to jerk away or "let go" from an electric shock. The danger in this situation is not from the current passing through the body, but from the reaction itself. Sometimes we let go in a manner that causes physical injury, such as cuts, bruises, or falls. At DC, the maximum current above which voluntary control is lost is approximately 60 mA for men and 35 mA for women. The let-go current follows a curve of the same shape as the perception current, reaching a minimum around 10 mA at 60 Hz before increasing again at higher frequencies. It is important to realize that these current levels are average, or typical, values; for some people the values for current may be lower.

The potentially lethal current level that could cause disruption of the electrical activity of the heart again varies with frequency and is a minimum at power-line frequencies. It is harder to quantify current levels because it is the amount of current that passes through

TABLE 9.1
Effects of electric current on the human body

Current	Physiological effect
1 mA	Perception of a faint tingling
5 mA	Slight shock felt. Disturbing, but not painful. Able to "let go." However, strong involuntary reaction or movement can cause injuries.
6–25 mA (women) 9–30 mA (men)	Painful shock. Loss of muscular control (i.e., "can't let go").
50–150 mA	Extremely painful shock, respiratory arrest (breathing stops), severe muscle contraction. Flexor muscles may cause holding on; extensor muscles may cause intense pushing away. Death is possible.
1,000–4,300 mA (1–4.3 A)	Ventricular fibrillation (disruption of rhythmic heart activity) occurs. Muscles contract; nerve damage occurs. Death is probable.
10,000 mA (10 A)	Cardiac arrest and severe burns occur. Death is probable.

the heart that is critical, and the amount of current passing through the heart depends on the entry and exit points of the current. At DC a current of 90 mA for men and 60 mA for women would probably cause serious injury or death, but at power-line frequencies, currents as low as 20–25 mA might cause injury or death.

At currents above 200 mA for men and 135 mA for women, tissue damage due to burning can occur. At these currents, there is enough heating in the tissue to cause irreversible changes in the tissue cells.

Avoiding electric shock means you have to avoid situations where electric current can flow through your body. Situations to avoid include using faulty or improperly connected equipment, damaged components, pinched wires, exposed conductors, and broken grounds. Care must also be used when operating equipment and components that require high voltages to function such as ionization gauges.

The body's first line of defense against current flow is the dry skin layer. This dry skin layer has a resistance in the range of 100,000 to 600,000 ohms at low frequencies. This large amount of resistance limits the amount of current that can flow into the body and out again. However, if the skin layer is moist due to perspiration, the resistance is significantly lower than dry skin and will allow a larger amount of current to flow into and out of the human body.

At RF frequencies and higher, the capacitance of the dry skin layer determines the impedance of the skin layer. The capacitance is in parallel with the resistance of the skin layer, creating two paths for current to flow into the human body. As frequency increases, the path through the capacitance of the skin layer becomes the preferred path for current.

When working around vacuum systems, there are a number of potential sources of electric shock and burns. Vacuum system components are powered by 120-volt, 208-volt, 220-volt, 230-volt, 240-volt, 380-volt, 480-volt, or other AC power sources. Motors and other system components require either single or three-phase power. In the United States, the line frequency is 60 Hz, but motors and other equipment are also designed to work at the 50-Hz line frequencies used in other countries.

Some vacuum system components require high DC voltages to operate. The Bayard-Alpert gauge is an example, requiring a voltage of 180 VDC to ground on the grid.

Here are safety precautions that should be followed when working around equipment:

- Be properly trained or qualified for the job.
- Plan the job and identify all safety issues.
- Never work alone. Use the "buddy" system.
- Isolate equipment from energy sources and follow required "lockout/tagout" procedures.
- Use required personal protective equipment, such as high-voltage gloves and rubber floor mats.
- Test every circuit and every conductor before you touch it.
- Use the "one hand" rule when testing circuits.

9.3 MECHANICAL HAZARDS

Mechanical hazards are produced by stationary and moving parts in and around vacuum systems and the larger processing system as a whole. Robotic arms, gears, pulleys, doors, and transport systems are just a few examples of moving parts that pose mechanical hazards.

The first step in avoiding mechanical hazards is to be aware of all the moving objects in your work area and to know the area in which they move. Safety guards are placed to prevent you from getting too near moving parts. In other situations, the operating perimeter of a moving object may be clearly marked to make you aware of the mechanical hazard. As a rule, never cross safety guards or enter the operating perimeter of a moving object.

Another step in avoiding mechanical hazards is to avoid wearing loose clothing or jewelry that may become entangled in moving equipment parts. If your clothing gets caught in a gear, you may be pulled into the machinery and suffer physical harm. Another serious situation can occur when neckties, scarves, or necklaces are caught in a mechanical mechanism, raising the possibility of strangulation. Or perhaps, if your clothing or jewelry is caught in a mechanical mechanism, you may not be able to move out of the way of other moving parts.

Mechanical hazards in process equipment using vacuum subsystems include load-lock doors that open and close, and the robotic arms that move product from cassettes into load locks and then back to the cassette.

9.4 VACUUM HAZARDS

Vacuum components such as gauges, chambers, piping can pose hazards if the integrity of the vacuum system is compromised. The potential hazards include implosion, explosion, exhaust problems, and chemical exposure.

Implosion occurs when the structure of a vacuum component is compromised or damaged. The force of the atmosphere will force the vacuum component to collapse upon itself, or implode. For example, if a glass vacuum gauge, such as a Bayard-Alpert gauge, cracks while under vacuum, the glass envelope of the gauge will implode. That is, the weakened glass barrier will crack under the force of the atmosphere, and the glass shards will be forced into the gauge by the rush of gas molecules. Then the shards will collide with each other and rebound in all directions. To provide protection should an implosion occur, glass gauges should be enclosed in a wire or metal shield.

Glass viewports pose the same kind of danger. Glass viewports are rated to below 10^{-11} torr. However, should the glass viewport be compromised, the atmosphere surrounding the vacuum system will force the glass port inward and litter the vacuum chamber with glass shards.

Explosions occur when the pressure inside the vacuum system is higher than the atmosphere surrounding it. Such explosions can occur when the vacuum system is pressurized, as when backfilling with nitrogen or other gas. If the pressure reaches a level above which a seal or valve is designed to operate, an explosion can occur. Explosions can also occur when two reactive gases are inadvertently mixed in the vacuum system, for example hydrogen gas mixed with oxygen gas or oxygen with a hydrocarbon.

Exhaust problems occur when the exhaust system is compromised. To detect exhaust problems, gas-monitoring systems are installed to detect gas leaks and alarms are triggered if a leak occurs. Leaks are especially hazardous when exhausting flammable and toxic gas by-products.

Cleaning vacuum chambers and components can pose chemical hazards. Residuals from manufacturing processes that are not removed by the vacuum system and remain in the chamber can pose a health risk. When cleaning a chamber or component, it is essential to know what materials you are cleaning. Use proper cleaning products and proper cleaning techniques. Wear required personal-protective equipment and follow basic chemical-safety rules.

9.5 THERMAL HAZARDS

Both hot and cold thermal hazards may exist in a processing system. Hot thermal hazards include motors, generators, heat exchangers, base plates, lamps, ion gauges, and ovens. All of these heat sources can cause serious burns when contact is made with them.

Cold thermal hazards are posed by the cryogenic liquids such as liquid nitrogen used in cold traps, baffles, and cryopumps. Liquid nitrogen, at 77 Kelvin, causes "cold burn," or frostbite, when it contacts the human body. Cryogenic liquids also pose an asphyxiation hazard. If allowed to evaporate into the atmosphere, the large amount of gas produced can deplete the oxygen level in the room, creating an oxygen-deficient environment.

When handling cryogenic liquids, special personal-protective equipment should be worn. This equipment includes insulating gloves and goggles with side shields or face shields.

Bakeout heaters raise the temperature in and around vacuum system components to promote the desorption of gas molecules from surfaces. This reduces the gas load produced by surface gas and results in lower pressures in the process chamber. Higher temperatures provide energy to break the bonds between the gas and vapor molecules adsorbed on the interior surfaces of the vacuum chamber and vacuum components attached to the chamber. Once the gas molecules desorb from the surface, they can be effectively removed from the chamber by the vacuum pumping subsystem.

Common bakeout temperatures are usually in the range of 200°C, but higher temperatures and longer bakeout times are used if pressures in the UHV range are desired. In this case, higher temperatures increase the diffusion rate of hydrogen and carbon monoxide in the stainless steel matrix so that these dissolved gases can migrate to the metal surface, desorb, and be pumped away.

Internal bakeout heaters are placed within the chamber to elevate the chamber temperature. Infrared lamps used for this purpose can raise the internal temperature of well-insulated chambers to 400°C. Higher internal temperatures, in the range of 900°C, can be achieved using quartz lamp heaters.

Vacuum system components generate heat under normal operation. Examples include hot ionization gauges and the variety of motors on vacuum pumps.

9.6 PRESSURIZED DEVICES

Pressurized devices include compressors, pneumatic systems, and compressed gas cylinders. All of these devices contain gas at a pressure higher than the atmosphere pressure in the operating environment.

Pneumatic systems are used to move product and to operate devices in a process tool. The amount of pressure required depends on the load requirements. The compressor and companion regulators are then used to produce the required gas pressure. When working on pneumatic lines and components, be aware of all high-pressure lines and components, and be sure to release the pressure prior to maintenance.

Another pressurized device found in a manufacturing facility is the compressed gas cylinder. These cylinders are pressurized to very high pressures, typically around 2,500 psig. Only personnel with specific training in the handling and safety of gas cylinders should handle them. Proper personal-protective equipment must be worn.

9.7 CHEMICAL HAZARDS

Training in chemical safety is an integral part of most new-employee orientation programs. Most follow the OSHA requirements contained in the Hazard Communications Standard (29 CFR 1910.1200).

Process chemicals and process by-products pose a wide range of health hazards to anyone in the manufacturing environment. These health hazards include irritation, sensitization, and asphyxiation. Chemicals also pose physical hazards, such as flammability and corrosion. It is imperative that technicians and other workers be cognizant of the chemicals they work around and the safety precautions that must be taken.

Some chemicals are carcinogens, which means that they are cancer producing. Others are toxic to the human body or can affect the reproductive system. Others can cause damage to sensitive membranes in the lungs and eyes. Exposure can be by inhalation, ingestion, and/or absorption, and can be acute (one-time exposure with an immediate or delayed reaction) or chronic (frequent exposures over a long period of time). The chemical's toxicity at the time of exposure depends on the length of exposure and the chemical's concentration. The permissible exposure limits are the concentration limits set by federal regulation and are spelled out in the chemical's Materials Safety Data Sheet (MSDS).

Each manufacturing process uses its own set of chemicals and thus poses its own set of chemical hazards. For example, processes using oxygen, such as reactive plasma cleaning and the reactive deposition of oxide compounds, can pose an explosion hazard. This can occur when pure oxygen is compressed in oil-sealed mechanical pumps, using hydrocarbon oil. Explosion risks can be minimized by using an oxygen-nitrogen mixture in place of pure oxygen, or using a pump oil that is more chemically stable, such as silicon oil or perfluorinated polyethers.

To minimize risk when using toxic gases such as arsine or flammable gases such as silane, gas distribution systems should be made of double-walled tubing. This type of piping allows any escaping gas to be trapped by the outer wall and pumped to a detection system where an alarm can be sounded.

Toxic gases can also be trapped within the pumping system. These gases can accumulate in pump oil, adsorb on interior surfaces, or react with other gases to form new toxic gases. Corrosive gases can cause deterioration in the vacuum system. For example, stainless steel surfaces are corroded by chlorine-containing plasma.

MSDS data sheets should be consulted for all chemicals used around or in process tools and systems, as well as for the reaction by-products that will be produced. Know the chemistry of the process that is being run.

SUMMARY

Safety cannot be taken lightly at any time. We humans are not invulnerable, and the soft tissue of our bodies can be damaged in an instant by chemical agents and physical hazards. Breathing, seeing, and hearing are complex mechanisms that can be irreparably damaged by these hazards.

Chemical and physical hazards can be minimized and avoided with awareness, discipline, and common sense. Safety rules and procedures are established to protect you, but you must follow them. Working with a buddy provides another set of eyes and ears to detect hazards and avoid them. Knowledge about safety increases your ability to see and avoid unsafe situations.

Safety is both an individual and a corporate responsibility. You must watch out for yourself and the people around you. Collective vigilance will keep us all safe.

BIBLIOGRAPHY

Brucher, Gerardo. "Prevention Is Key to Vacuum Systems," *R&D Magazine,* February 2001, p. 57.

Comello, Vic. "Exhausting Process Gases Safely and Efficiently," *R&D Magazine,* February 1998, p. 81.

Mattox, Donald M. "Safety Aspects of Vacuum Processing," *Vacuum Technology & Coating Magazine,* January 2005, p. 32–34.

INDEX